I0051373

Mastering Prometheus

Gain expert tips to monitoring your infrastructure, applications, and services

William Hegedus

‹packt›

Mastering Prometheus

Copyright © 2024 Packt Publishing

All rights reserved. No part of this book may be reproduced, stored in a retrieval system, or transmitted in any form or by any means, without the prior written permission of the publisher, except in the case of brief quotations embedded in critical articles or reviews.

Every effort has been made in the preparation of this book to ensure the accuracy of the information presented. However, the information contained in this book is sold without warranty, either express or implied. Neither the author, nor Packt Publishing or its dealers and distributors, will be held liable for any damages caused or alleged to have been caused directly or indirectly by this book.

Packt Publishing has endeavored to provide trademark information about all of the companies and products mentioned in this book by the appropriate use of capitals. However, Packt Publishing cannot guarantee the accuracy of this information.

Group Product Manager: Preet Ahuja

Publishing Product Manager: Suwarna Rajput

Book Project Manager: Ashwin Kharwa

Senior Editor: Sayali Pingale

Technical Editor: Arjun Varma

Copy Editor: Safis Editing

Proofreader: Safis Editing

Indexer: Tejal Daruwale Soni

Production Designer: Nilesh Mohite

DevRel Marketing Coordinator: Rohan Dobhal

First published: May 2024

Production reference: 1220324

Published by Packt Publishing Ltd.

Grosvenor House

11 St Paul's Square

Birmingham

B3 1RB, UK

ISBN 978-1-80512-566-2

www.packtpub.com

To Curtis John, for getting me started in my career, supporting me professionally and personally, and being willing to let me boss him around at multiple companies. To my wife, Kaylah, for her sacrifices, encouragement, love, and grace throughout the process of writing this book and all my life. Finally, to my sons, Luke, Joel, Zeppelin, and Chara, for always keeping my priorities straight.

– William Hegedus

Contributors

About the author

William Hegedus has worked in tech for over a decade in a variety of roles, culminating in site reliability engineering. He developed a keen interest in Prometheus and observability technologies during his time managing a 24/7 NOC environment and eventually became the first SRE at Linode, one of the foremost independent cloud providers.

Linode was acquired by Akamai Technologies in 2022, and now Will manages a team of SREs focused on building the internal observability platform for Akamai's Connected Cloud. His team is responsible for a global fleet of Prometheus servers spanning over two dozen data centers and ingesting millions of data points every second, in addition to operating a suite of other observability tools.

Will is an open source advocate and contributor who has contributed code to Prometheus, Thanos, and many other CNCF projects related to Kubernetes and observability. He lives in central Virginia with his wonderful wife, four kids, three cats, two dogs, and a bearded dragon.

I want to thank my wife and kids for their support and understanding through the late nights and busy weekends of writing this book. Additionally, I want to thank the Prometheus community of maintainers and contributors who are so welcoming and helpful – with special thanks to Ben Kochie and Bartek Plotka, who were so influential in my early learning of Prometheus and Thanos. Finally, I want to thank my team at Akamai for being so supportive and encouraging and being my inspiration for writing this book.

About the reviewer

TJ Hoplock is an experienced SRE who specializes in observability. He was an early adopter of Prometheus while working for cloud hosting company Linode (now Akamai Connected Cloud) and a founding member of Linode's Observability team. He is an active contributor to Prometheus (contributing to Linode service discovery, among other improvements), Alertmanager, and other open source tools. He was previously a technical reviewer for *Prometheus: Up and Running, 2nd Edition* by Julien Pivotto. He is currently working as a senior SRE at NS1 (an IBM company), playing a leading role working with tools such as OpenTelemetry, Prometheus, Thanos, Loki, Honeycomb, and Orb to improve NS1's observability posture and improve services for users.

Table of Contents

Part 2: Scaling Prometheus

6

Advancing Prometheus: Sharding, Federation, and High Availability   95

7

Optimizing and Debugging Prometheus   111

8

Enabling Systems Monitoring with the Node Exporter 131

Part 3: Extending Prometheus

9

Utilizing Remote Storage Systems with Prometheus 147

10

Extending Prometheus Globally with Thanos 167

11

Jsonnet and Monitoring Mixins   193

12

Utilizing Continuous Integration (CI) Pipelines with Prometheus   225

Preface

Since the Prometheus project was first announced to the world in January 2015, it has rapidly become the de facto modern monitoring solution. Open source projects such as Kubernetes expose Prometheus metrics by default, cloud providers sell "managed" Prometheus services, and it even has its own yearly conference. However, in my personal journey to learn and understand Prometheus deeper, I came to a saddening realization. There are a plethora of blog posts, books, and tutorials focused on the basics of Prometheus, but few to no readily available resources that cover running Prometheus at scale.

For that information, I found myself needing to turn to conference talks or trying to extrapolate how others do it by reading through GitHub issues on the Prometheus repository, questions on the Prometheus mailing list, and conversations in the official Prometheus Slack channel. Those learnings – coupled with years of personal experience – have gone into this book as an endeavor to begin filling that void.

Perhaps one of the limiting factors of existing content is that it tends to focus purely on Prometheus itself, omitting the larger ecosystem surrounding it. However, to run Prometheus at scale, it quickly becomes necessary to build on top of Prometheus to extend it. Instead of Prometheus being the destination, it provides a foundation.

We will still cover Prometheus itself in depth. We will see how to get the most out of it, implement best practices, and – perhaps most critically – develop a deeper understanding of how Prometheus's internals work.

In addition to Prometheus itself, though, we'll also look at how to operate and scale Prometheus. We'll learn how to debug Prometheus through Go's developer tools, how to manage hundreds (or thousands) of Prometheus rules without losing your mind, how to connect Prometheus to remote storage solutions, how to run dozens of highly available Prometheus instances while maintaining a global query view, and much, much more.

Who this book is for

This book is primarily intended for readers with a preexisting, basic knowledge of Prometheus. You're probably already running Prometheus, know what the Node Exporter and Alertmanager are, and can write PromQL queries without too much help.

If this is your first introduction to Prometheus, that's OK, too! The content should still be generally accessible, but you may – at times – wish to seek out additional resources to explain some of the basics that are glossed over.

Regardless of your experience level, this book is targeted at *operators* of Prometheus. If you're a developer who uses a Prometheus environment operated by someone else, there are still chapters that will be of interest to you, but there are others that are unlikely to be applicable to your responsibilities.

What this book covers

Chapter 1, Observability, Monitoring, and Prometheus, gives a brief overview of the history of modern monitoring systems, establishes common observability terminology and concepts, and looks at Prometheus' role within observability.

Chapter 2, Deploying Prometheus, goes through the process of deploying Prometheus to Kubernetes and provides the lab environment that we will use throughout the rest of the book.

Chapter 3, The Prometheus Data Model and PromQL, dives deep into the technical specifics of how Prometheus – and especially its **Time Series DataBase** (**TSDB**) – works, along with an overview of how the **Prometheus Query Language** (**PromQL**) works.

Chapter 4, Using Service Discovery, goes into the details of how to use dynamic service discovery in Prometheus, including how to build your own service discovery providers.

Chapter 5, Effective Alerting with Prometheus, focuses on making Prometheus alerting reliable and testable, along with making the most of the Alertmanager.

Chapter 6, Advancing Prometheus: Sharding, Federation, and HA, is where we begin to look at scaling Prometheus past a single Prometheus server and into reliable, distributed deployments.

Chapter 7, Optimizing and Debugging Prometheus, explores how to leverage Go tools to debug the Prometheus application and how to tune Prometheus for optimum performance.

Chapter 8, Enabling Systems Monitoring with the Node Exporter, looks in depth at the most commonly deployed Prometheus exporter to understand all that it can do.

Chapter 9, Utilizing Remote Storage Systems with Prometheus, examines Grafana Mimir and VictoriaMetrics as two options that Prometheus can send data to for long-term storage, global query view, multi-tenancy support, and more.

Chapter 10, Extending Prometheus Globally with Thanos, comprehensively explores all components of the Thanos project to see how they can be used to extend the functionality of Prometheus, enable high availability, and provide for nearly unlimited retention of metrics.

Chapter 11, Jsonnet and Monitoring Mixins, introduces the Jsonnet programming language as a tool to simplify the management of Prometheus rules at scale. Additionally, we see how the Monitoring Mixins project from various Prometheus maintainers and contributors makes use of Jsonnet to provide configurable, reusable alerts and dashboards for various systems and software.

Chapter 12, Utilizing Continuous Integration (CI) Pipelines with Prometheus, takes a practical look at how you can manage your Prometheus configuration and alerts in Git and perform a variety of automated tests to them to ensure they are valid and conform to expectations.

Chapter 13, Defining and Alerting on SLOs, explores how Prometheus can be used to define, measure, and alert on **Service Level Objectives** (**SLOs**), including through the use of open source tools such as Pyrra and Sloth that make it easy to implement best-practice SLO alerting.

Chapter 14, Integrating Prometheus with OpenTelemetry, takes a look at the OpenTelemetry project – its history, its future, and how Prometheus integrates with it.

Chapter 15, Beyond Prometheus, brings us full circle back to our initial discussion about observability and provides ideas on where to go next in building out your observability suite.

To get the most out of this book

You will need to have some level of hands-on experience with Prometheus, a basic understanding of server administration, and – ideally – some prior experience with Kubernetes. The lab environment used throughout this book is based on Kubernetes, but all commands used to interact with Kubernetes are explicitly stated. Therefore, prior Kubernetes knowledge is not required, but it will certainly aid in deeper understanding and enable further experimentation outside of what is covered in the book.

Software/hardware covered in the book	Operating system requirements
Prometheus	Windows, macOS, or Linux
Thanos	Windows, macOS, or Linux
OpenTelemetry	Windows, macOS, or Linux
Kubernetes	Windows, macOS, or Linux

If you are using the digital version of this book, we advise you to type the code yourself or access the code from the book's GitHub repository (a link is available in the next section). Doing so will help you avoid any potential errors related to the copying and pasting of code.

Download the example code files

You can download the example code files for this book from GitHub at `https://github.com/PacktPublishing/Mastering-Prometheus`. If there's an update to the code, it will be updated in the GitHub repository.

We also have other code bundles from our rich catalog of books and videos available at `https://github.com/PacktPublishing/`. Check them out!

Conventions used

There are a number of text conventions used throughout this book.

`Code in text`: Indicates code words in text, database table names, folder names, filenames, file extensions, pathnames, dummy URLs, user input, and Twitter handles. Here is an example: "Mount the downloaded `WebStorm-10*.dmg` disk image file as another disk in your system."

A block of code is set as follows:

```
>>> hashB = int(md5(SEPARATOR.join(targetB).encode("utf-8")).
hexdigest(), 16)
>>> hashB
139861250730998106692854767707986305935

>>> print(f"{targetA} % {MOD} = ", hashA % MOD)
['app=nginx', 'instance=node2'] % 2 =  0

>>> print(f"{targetB} % {MOD} = ", hashB % MOD)
['app=nginx', 'instance=node23'] % 2 =  1
```

When we wish to draw your attention to a particular part of a code block, the relevant lines or items are set in bold:

```
alerting:
  alert_relabel_configs:
    - regex: prometheus_replica
      action: labeldrop
```

Bold: Indicates a new term, an important word, or words that you see onscreen. For instance, words in menus or dialog boxes appear in **bold**. Here is an example: "Select **System info** from the **Administration** panel."

> **Tips or important notes**
> Appear like this.

Get in touch

Feedback from our readers is always welcome.

General feedback: If you have questions about any aspect of this book, email us at `customercare@packtpub.com` and mention the book title in the subject of your message.

Errata: Although we have taken every care to ensure the accuracy of our content, mistakes do happen. If you have found a mistake in this book, we would be grateful if you would report this to us. Please visit www.packtpub.com/support/errata and fill in the form.

Piracy: If you come across any illegal copies of our works in any form on the internet, we would be grateful if you would provide us with the location address or website name. Please contact us at copyright@packt.com with a link to the material.

If you are interested in becoming an author: If there is a topic that you have expertise in and you are interested in either writing or contributing to a book, please visit authors.packtpub.com.

Share Your Thoughts

Once you've read *Mastering Prometheus*, we'd love to hear your thoughts! Scan the QR code below to go straight to the Amazon review page for this book and share your feedback.

https://packt.link/r/1-805-12566-4

Your review is important to us and the tech community and will help us make sure we're delivering excellent quality content.

Download a free PDF copy of this book

Thanks for purchasing this book!

Do you like to read on the go but are unable to carry your print books everywhere?

Is your eBook purchase not compatible with the device of your choice?

Don't worry, now with every Packt book you get a DRM-free PDF version of that book at no cost.

Read anywhere, any place, on any device. Search, copy, and paste code from your favorite technical books directly into your application.

The perks don't stop there, you can get exclusive access to discounts, newsletters, and great free content in your inbox daily

Follow these simple steps to get the benefits:

1. Scan the QR code or visit the link below

https://packt.link/free-ebook/978-1-80512-566-2

2. Submit your proof of purchase
3. That's it! We'll send your free PDF and other benefits to your email directly

Part 1: Fundamentals of Prometheus

In this part, we'll get acquainted with observability concepts and understand the role that Prometheus serves in adding observability to your systems and services. Then, we'll dive into Prometheus itself by deploying it and going through a detailed examination of its data model and query language. Finally, we'll see how to leverage built-in Prometheus service discovery features to make the configuration of scrape targets dynamic, and how to make the most of its alerting features.

This part has the following chapters:

- *Chapter 1, Observability, Monitoring, and Prometheus*
- *Chapter 2, Deploying Prometheus*
- *Chapter 3, The Prometheus Data Model and PromQL*
- *Chapter 4, Using Service Discovery*
- *Chapter 5, Effective Alerting with Prometheus*

1

Observability, Monitoring, and Prometheus

Observability and monitoring are two words that are often used synonymously but carry important distinctions. While this book is not focused on academic definitions and theories surrounding observability, it's still useful to distinguish between observability and monitoring because it will provide you with a framework to get in the right mindset when thinking about how Prometheus works and what problems it solves. A screw and a nail can both hang a picture, and you can bang a screw into a wall with a hammer, but that doesn't make it the best tool for the job. Likewise, with Prometheus, I've seen many people fall into the trap of trying to use Prometheus to cover all of their observability and monitoring needs – when you have a hammer, everything looks like a nail. Instead, let's identify where Prometheus shines so that we can use it to its full effect throughout the rest of this book.

In this chapter, we're going to cover the following main topics:

- A brief history of monitoring
- Introduction to observability concepts
- Prometheus's role in observability

Let's get started!

A brief history of monitoring

In the beginning, there was Nagios… or, at least, so the story goes. Monitoring as we know it took off in the late 1990s and early 2000s with the introduction of tools such as Nagios, Cacti, and Zabbix. Sure, some things existed before that that focused on network monitoring such as **Multi Router Traffic Grapher** (**MRTG**) and its offshoot, rrdtool, but system monitoring – including servers – found its stride with Nagios. And it was good… for a time.

Nagios (and its ilk) served its purpose and – if your experience is anything like mine – it just won't seem to go away. That's because it does a simple job, and it does it fairly well. Let's look a little closer at it, the philosophy it embodies, and where it differs from Prometheus.

Nagios

Early monitoring tools such as Nagios were **check-based**. You give it a script to run with some basic logic and it tells you whether things are good, bad, or *really* bad based on its exit code. This philosophy defined early monitoring – there wasn't much nuance to it. Even more so, there wasn't much *detail* to it.

In the early days of monitoring as a discipline, systems administrators were primarily focused on **known unknowns**. In other words, you'd set up checks for how you'd expect your system to break. High memory utilization? Got a check for that. CPU usage high? Got a check for that. Process stopped running? You know I've got a check for that.

This has obvious limitations: namely, you need to know in advance what ways your systems can and will break. It was primarily focused on **cause-based** alerting, in which you set up an alert for something that you know will cause a problem (get it?). However, Nagios and monitoring tools like it also differed in another key way: they assumed a fairly static set of monitoring targets.

Before the cloud, before virtualization took off, there was cold, hard bare-metal. Well, probably not cold if it's doing any work, but you get the picture. Buying and provisioning servers was a far cry from the modern cloud-based experience we enjoy now, in which you can have a server running the latest release of Ubuntu halfway across the world created and booted in less than a minute. Consequently, the monitoring tools of the time didn't have to worry about constantly changing, dynamic environments. Nor did they have to account for servers running across the globe. Adding or removing monitoring targets and checks was pretty infrequent, and you didn't need a system that could do fancy stuff such as continuous, dynamic service discovery. You could tolerate a blip in monitoring while Nagios fully restarted to pick up a new configuration (something it still must do to this day).

Nevertheless, Nagios and other tools didn't *need* that functionality. They weren't pain points the way they are today. Nagios didn't make a system "observable" – it just monitored if it was healthy or broken, and that was *fine*. For a time.

A word on SNMP

While Nagios was the first big tool to focus on systems monitoring, network monitoring tools existed for at least a decade before it. The first RFCs for defining **Simple Network Management Protocol (SNMP)** were submitted in 1988, and it is still in use widely today.

Originally, SNMP was designed primarily as a protocol to manage network devices rather than to monitor them. These days, though, SNMP is so synonymous with network monitoring that many people assume that the M in SNMP stands for monitoring. It can be used for things such as monitoring how many packets a particular network interface has or the chassis temperature of a switch. It's the default way to monitor most networking equipment on the market and Prometheus is not a replacement for it. One of the most popular Prometheus exporters that is maintained by the core Prometheus team is the SNMP exporter.

Enter the cloud

While Nagios and similar tools were in their heyday in the 2000s, a disruptive new landscape was forming that we'd come to call cloud computing. Multiple advances in virtualization technologies such as Xen and KVM helped to enable the cloud as we know it today. At this point, it was possible and accessible to run virtual servers across the globe, which meant no more dealing with lead times to order and receive physical hardware and no more being physically restricted to nearby co-located data centers/the office basement/a spare closet. This marked a major challenge to monitoring as it had thus far been done.

As we previously covered, Nagios – and the tools developed around the same time – was (and is) well suited to static environments. However, with the cloud coming into its own, engineers started to realize that it took as long (or longer) to restart Nagios to pick up a new config as it did to create a new **virtual machine** (**VM**) from scratch. This wasn't scalable.

Additionally, as systems became more complex – monoliths were broken into microservices, servers were created and destroyed as load ebbed and flowed – the need to have better observability into how these distributed systems were performing increased. No longer could a simple HTTP check suffice to know if the user experience of your site was good – the site was no longer just a single, carefully managed server. We needed to know things such as what our error rate looked like across all of our backends. Once you have that, you want to know if it's gone up or down since the latest release was deployed. You know how it goes: if you give a mouse a cookie, they're going to ask for a glass of milk. Simple monitoring couldn't cut it anymore in these changing conditions – we needed observability.

Introduction to observability concepts

Observability both as a word and as a discipline is not unique to technology. The term is derived from control theory, which is traditionally more rooted in physical engineering disciplines such as robotics and nuclear engineering. It is, in essence, the ability to surmise the health of a system by observing its inputs and outputs. In nuclear engineering, you put in uranium and water, and you receive heat and steam. In software engineering, you put in an end user and an API call, and you receive a Jira ticket about how your API isn't working. Err… well, hopefully not if your observability is doing its job.

Observability in systems engineering and software is primarily informed by and achieved with a handful of important telemetry signal types. You may have heard them referred to as "the three pillars of observability," but that terminology has since fallen out of fashion as it elevates the act of gathering telemetry itself above the result. Make no mistake, you can have all three pillars without having an observable system. So, what are these pillars anyway? Well, they're **metrics**, **logs**, and **traces**.

> **Other telemetry types**
>
> This does not preclude other forms of telemetry, though! For example, since 2020, there has been an explosion of exciting new technologies around **continuous profiling** through companies such as Polar Signals and Pyroscope (since acquired by Grafana Labs). Additionally, there is a line of thinking championed by observability sages such as Charity Majors, who argue that **events** are the foundational building blocks of observability.

Metrics

Metrics are perhaps the most fundamental type of telemetry. They are what most organizations start with in their observability journey. The reason is two-fold: they are cheap (in Prometheus, each data point for a metric takes up less than 2 bytes on average) and they are simple. Additionally, metrics are core to the time series-based alerting we mentioned earlier. So, what is a metric?

A metric is a numeric, time-based representation of one or more scoped data points. For example, you may have a metric on memory usage and watch how that changes over time, or a metric on error rates returned by an application. Metrics don't *have* to be time series (though they often are), but they are always related to time since they have a relationship with the time in which they were observed/recorded.

They can also be viewed as statistics. In the business world, **key performance indicators** (**KPIs**) are also measured via metrics. A lot of advancements in the SRE space at the time of writing have to do with treating time series metrics as statistics and applying concepts from the mathematical branch of statistics to them to analyze distributions and compute z-scores (a novel way to detect anomalies by scoring how much of an outlier a data point is).

Concerning Prometheus, metrics are the primary observability signal that it is concerned with. Prometheus is purpose-built as a time series database for storing metric data efficiently.

Time series

If a time series is different from a metric, then how can you tell the difference between the two? The simplest explanation involves an example.

Recall when I mentioned tracking memory usage as a metric. Well, this is how that looks in Prometheus with a metric coming from the `node_exporter` process:

```
node_memory_MemFree_bytes
```

Each metric can have one or more **time series** associated with it. Each time series is an instance of a metric. If you're familiar with object-oriented programming, you can think of a metric as a base class that multiple time series inherit from. Individual time series distinguish themselves through different metadata (that is, "labels") attached to them. Consequently, individual time series in Prometheus look like this:

```
node_memory_MemFree_bytes{instance="server1.example.com"}
node_memory_MemFree_bytes{instance="server2.example.com"}
```

Notice how the metric name is still the same, but the labels (in the curly braces) differ. For every metric, you can have tens, hundreds, or even thousands of individual time series for it (although this relates to cardinality, which we will discuss later).

Metrics, however, are limited in their amount of detail. By nature, many are aggregations. For example, counting the number of requests your API receives in total – not what each request looked like in terms of user-agent header, IP address, and so on. Consequently, you trade some level of detail in exchange for greater efficiency in querying and storage. However, the next two telemetry signals we'll discuss regain some of that lost detail and are also critical to an observable system.

Logs

Logs are the OG telemetry signal. Before terminal screens existed, computers were *printing* log lines. They are the easiest to implement in a vacuum and are often quick-and-dirty (anyone else fall back on using `printf` to debug an issue instead of an actual debugger?). However, they are also inherently unstructured. Logs are typically meant to be read by humans rather than machines. Humans don't want to read flattened JSON; we're much more likely to add a long line that looks like, HEY!!! LOOK AT ME!! ERROR: `<stacktrace>`. Consequently, logs can be the most difficult signal to extract meaningful value from.

Logs are simply any bit of text emitted by a system designed to produce a record of something occurring. There are some common types of logs, such as access logs from Apache or Nginx, that have a consistent format. However, most logging is arbitrarily formatted. To maximize the usefulness of logs for observability, logs ought to be structured, which is a fancy way of saying that they should conform to a predictable format. To many people, that means writing out logs in JSON. To folks in the Go community, it's often `logfmt` (what Prometheus uses). Regardless, the important piece is that structured formats provide a way to predictably extract **key/value (K/V)** pairs from log lines. These K/V pairs can then be used as a dimension upon which aggregations can be performed to identify trends and anomalies.

For comparison's sake, here's an example of a structured log from Prometheus using `logfmt`:

```
ts=2023-06-14T03:00:06.215Z caller=head.go:1191 level=info
component=tsdb msg="WAL checkpoint complete" first=1059 last=1060
duration=1.949831517s
```

In a sense, logs are the closest thing that most companies have to what is referred to by some in the observability space as "events." Events should be "arbitrarily wide," as the aforementioned Charity Majors likes to say, which is to say that they should be able to have as many or as few K/V pairs as necessary without it mattering much. Structured logs fit that bill.

Traces

Traces are often the last observability signal to be adopted by organizations. It's the least distorted signal, but also the most cumbersome to implement. An application trace can be thought of as a sort of custom stack trace that only tracks functions or steps in your code path that you care about. Each trace is made up of one or more **spans**, which represent individual units of work (often functions):

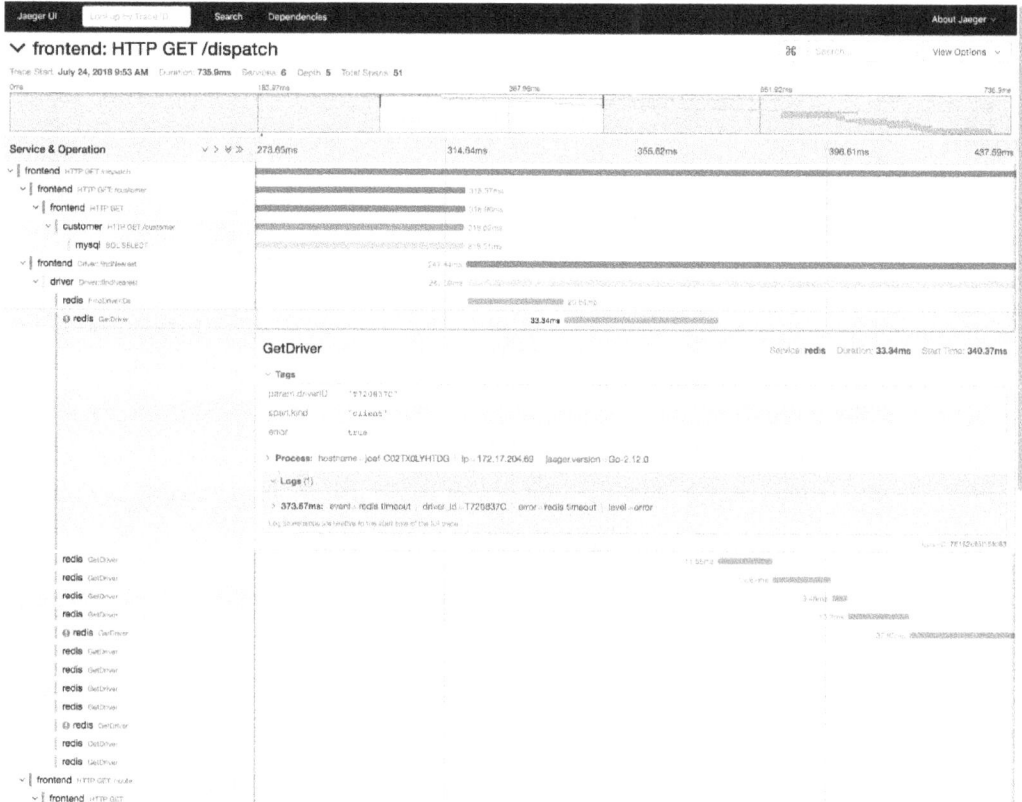

Figure 1.1 – An example of a trace with multiple spans being visualized in the open source Jaeger project (`https://github.com/jaegertracing/jaeger`)

A special but common type of trace is a **distributed trace**, which is a trace that encompasses multiple systems as a request travels to those distinct systems all while maintaining the same trace ID. It is an SRE's dream – to be able to follow a request from the moment it hits your public load balancer to the time it hits your database from the 12th microservice it called.

However, all this wonderful detail is not without its drawbacks. Tracing is *expensive*. There is overhead from generating the trace, the cost to store and retrieve this detailed data, and the cost of running the tracing infrastructure itself (which is often far more involved than running something such as Prometheus). Consequently, tracing data is often *sampled* to reduce those costs. Depending on the workload and its expected requests per second, I've generally seen 1/100 as the sample rate for many production applications. This means that for every 100 requests, only 1 trace is stored. This can be problematic in theory since you may be missing 99 failed requests and only seeing the 1 successful request, or vice versa. Nevertheless, it's typically sufficient to provide insight.

> **Prometheus and tracing**
>
> A neat, relatively new feature of Prometheus called **exemplars** allows Prometheus to integrate with tracing tools by attaching a trace ID to time series data points. This enables users to quickly jump to a trace from around the same time as a spike in a graph of your Prometheus data, which can be time-saving and is another way to illustrate the symbiotic connection between the various observability signals.

Other signals

Metrics, logs, and traces are not the be-all-and-end-all of observability – their mere presence does not make a system observable. Additionally, even more observability signals are emerging to continue to build a more complete picture of systems and thereby make them more observable. The two with the most traction behind them are events and profiles.

Events

We've touched briefly on events already, but to recap, events are point-in-time records of an action occurring. These *could* be logs (see the article on canonical log lines in the *Further reading* section), but they could also be purpose-built systems or even a database table or two. An event, however, can also be viewed as an all-encompassing signal that includes metrics, logs, and traces. An event may include a server's memory and CPU utilization at the time of its generation (metrics), a tack on a field containing logs generated through the duration of the request (logs), and even information about a unique request ID/trace ID.

At the time of writing, there are no great open source observability tools for events (that I'm aware of), but plenty of SaaS vendors such as Honeycomb and New Relic are leading the charge in both defining and supporting this data type.

Profiling

Profiling has been around for a while, but only recently has it become a more prevalent observability signal in the form of **continuous profiling**. Profiling has long been a developer tool for analyzing application and system performance (for example, kprobes have been present in the mainline Linux kernel since as far back as 2005). However, it was such an intrusive, interruptive process that it was only used as needed. Now, almost two decades later, advances in computing have made it possible to profile a system continuously instead of just as needed.

This profiling data often looks like a trace, which makes sense – profiling involves running stack traces, and application traces are similar to stack traces. They are visualized in the same way (flame graphs) and stored similarly.

Unlike events, several open source continuous profiling tools exist, including Parca, from Polar Signals, a company started by a long-time core contributor to Prometheus, Thanos, and many other Prometheus-related projects.

Tying signals together

An observable system cannot be observable from only one point of view. A computer monitor that is only viewable from directly in front of it at a 90-degree angle to the screen would not sell well, and similarly, your systems should be observable from multiple angles. Going all in on logs without metrics or traces leaves gaps and inefficiencies in your observability. The same logic goes for any one signal. Instead, they all build off each other.

I like to think of the relationship between observability signals as different levels of drilling down into a problem. I start with metrics – they show trends and alert me when something is out of the ordinary. Then, I'll look at the logs for a system. Ideally, the logs have a unique request ID attached to them, which I can then use to jump to look at the trace for an individual request and see where the issue occurred. From there, you can use the more detailed trace data to try to ascertain trends, such as "all failed requests went through this load balancer." Each signal enables you to dig deeper into the next and in a more targeted manner.

Getting data out of systems

One of the core features that distinguishes Prometheus from Graphite, Nagios (when using passive checks), and most other monitoring systems is that it uses a **pull-based** model for extracting data from the systems it monitors. This contrasts with other systems that use a **push-based** model, in which you push metrics to a destination.

Prometheus constantly pulls metrics from targets, which offers several advantages. A major advantage is that Prometheus knows what it is supposed to be monitoring and can easily alert you to any issues it encounters with monitoring those things. This makes silent failures far less likely to occur compared to a push-based monitoring system where you not only need to track what the values of metrics are but also how recently they were been updated.

We've mentioned how Prometheus focuses on metrics, but where does Prometheus itself fit into observability as a whole? Let's take a look.

Prometheus's role in observability

Prometheus is objectively pretty great at what it does, but can we make a system fully observable with just Prometheus? Unfortunately, the answer to that question is no. Prometheus's strength is also its weakness – it's singularly focused on one thing: metrics.

Prometheus is only focused on the metrics aspect of an observable system. It is purpose-built to efficiently store numeric time series of varying types in a simple format. To the extent to which it interoperates with other observability signals, it is only to provide a link or bridge to some other purpose-built system.

However, Prometheus also provides some of the highest-value data that you can collect. It's likely your go-to data source in tools such as Grafana to visualize how your systems are performing. Logging and tracing systems can certainly provide more detailed data, but to get them to provide that same level of value in analyzing trends, you need to strip *away* detail to aggregate data. Consequently, Prometheus is often going to be your best bet for alerting and dashboarding.

Alerting

Prometheus shines in the area of alerting. Alerting has to do more specifically with **monitoring** versus observability, but monitoring is inextricably tied to observability – the more observable your system is, the better the monitoring can be.

Monitoring can be thought of as the discipline of proactively identifying and alerting on the breaching of predetermined thresholds. Prometheus, by nature of being a time series database, excels in being able to alert on trends over time. When comparing Prometheus to Nagios, often, the first contrast I point to is just how much better Prometheus is at alerting.

It's not some complex feature that makes Prometheus better than check-based monitoring and alerting software such as Nagios. It's fairly simple. To me, Prometheus's killer feature in alerting is just the ability to specify a **for** duration in alerts. For an alert to go from a *pending* to a *firing* state, it needs to meet the alerting criteria consistently across the length of a *for* duration.

While Nagios and other monitoring tools like it have the idea of hard and soft states and check retries, you can still get unlucky with check timing. Maybe Nagios just happens to check CPU usage every 5 minutes when a bursty cron job runs. With Prometheus, you're almost certainly getting those metrics way more frequently than every 5 minutes, which means that you're getting more comprehensive data more frequently, which leads to a more accurate picture of how a system is performing. All this leads to more reliable alerting from Prometheus with less "flappy" and noisy alerts, leading to less alert fatigue and better reliability.

Dashboarding

Dashboarding is arguably what people think of first when they think of observability. "An observable system is one where I can see a bunch of lines that go up and down." There's plenty of discourse on the value of dashboards overall, and certainly jumping from dashboard to dashboard to try to find the source of a problem is not ideal observability. Nevertheless, dashboards are the darling child of executives, SREs, and developers alike. They're simple, intuitive, and easy to set up once you have the data. And where does that data come from? Why, Prometheus of course!

Prometheus is the most used data source plugin for the most popular open source dashboarding tool out there – Grafana. Because Prometheus is generally more high-level than logs or traces – since it deals with data in aggregate rather than per request – it can be the most useful data source to produce graphs and use in dashboards.

What Prometheus is not

We've covered how Prometheus directly relates to the signal type of metrics, but it's worth mentioning a few specific things that Prometheus is explicitly *not*:

- Prometheus is *not* an all-in-one monitoring solution
- Prometheus is *not* something you configure once and forget about
- Prometheus is *not* going to automatically make your systems more observable

In fact, in terms of what telemetry signal provides the most detailed view of your system, metrics are the least specific. So, is it even worth running Prometheus? (Yes.) Should I just go all in tracing and forget the rest? (No.) Metrics – and, by extension, Prometheus – are pivotal to monitoring and can still contribute to making a system more observable (especially with custom instrumentation)! Throughout the rest of this book, we'll look at how to use Prometheus to its maximum effect to make all your systems more observable.

Summary

In this chapter, we learned the abridged history of monitoring and observability, what observability is, and how Prometheus contributes to observability through metrics. With this new (or refreshed) frame of mind, we can approach our utilization of Prometheus in a way that maximizes its usefulness without trying to use it as a silver bullet to solve all our problems.

In the next chapter, we'll cover deploying a Prometheus environment that we'll use as the foundation that we build upon throughout the remainder of this book.

Further reading

To learn more about the topics that were covered in this chapter, take a look at the following resources:

- *Fast and flexible observability with canonical log lines*: `https://stripe.com/blog/canonical-log-lines`

2

Deploying Prometheus

Now that we understand a little more about observability and monitoring, as well as their distinctions and how Prometheus fits into it all, we can deploy Prometheus. In this book, we'll be deploying Prometheus to a managed Kubernetes environment using **Linode Kubernetes Engine** (**LKE**). However, you can substitute AWS's EKS, Azure's AKS, or any other engine as you see fit.

You may already have experience with deploying Prometheus, but we're going to use the Prometheus environment that we'll deploy in this chapter as the basis for future chapters in which we'll be deploying additional applications that extend Prometheus and make configuration changes to Prometheus. Consequently, it is recommended that you follow along with the deployment, even if you have prior experience.

In this chapter, we will cover the following topics:

- Components of a Prometheus stack
- Provisioning Kubernetes
- Deploying the Prometheus Operator

Let's get started!

Technical requirements

You'll need to install the following programs and make them available in your PATH:

- **linode-cli**: `https://github.com/linode/linode-cli`
- **kubectl**: `https://kubernetes.io/docs/tasks/tools/#kubectl`
- **helm**: `https://helm.sh/docs/intro/install/`

This chapter's code is available at `https://github.com/PacktPublishing/Mastering-Prometheus`.

Components of a Prometheus stack

When discussing Prometheus, people rarely mean just the Prometheus project (`https://github.com/prometheus/prometheus`) itself. Instead, they refer to the following:

- Prometheus as a full solution in which you deploy Prometheus for metrics collection, storage, and alerting rule evaluation

- Prometheus's Alertmanager (`https://github.com/prometheus/alertmanager`) for routing alerts coming from Prometheus

- One or more exporters, such as Node Exporter (`https://github.com/prometheus/node_exporter`), to expose metrics

- Grafana (`https://github.com/grafana/grafana`) to visualize metrics from Prometheus via dashboards

Each of these components fits together to form a complete metrics solution for observability.

Prometheus

Understandably, Prometheus is the core technology in a Prometheus stack. Prometheus is comprised of four main parts: the **time series database** (**TSDB**), the **scrape manager,** the **rule manager**, and the **web UI** (along with its REST API). There are – of course – additional components, as shown in the following figure, but these are the main ones you need to be concerned with:

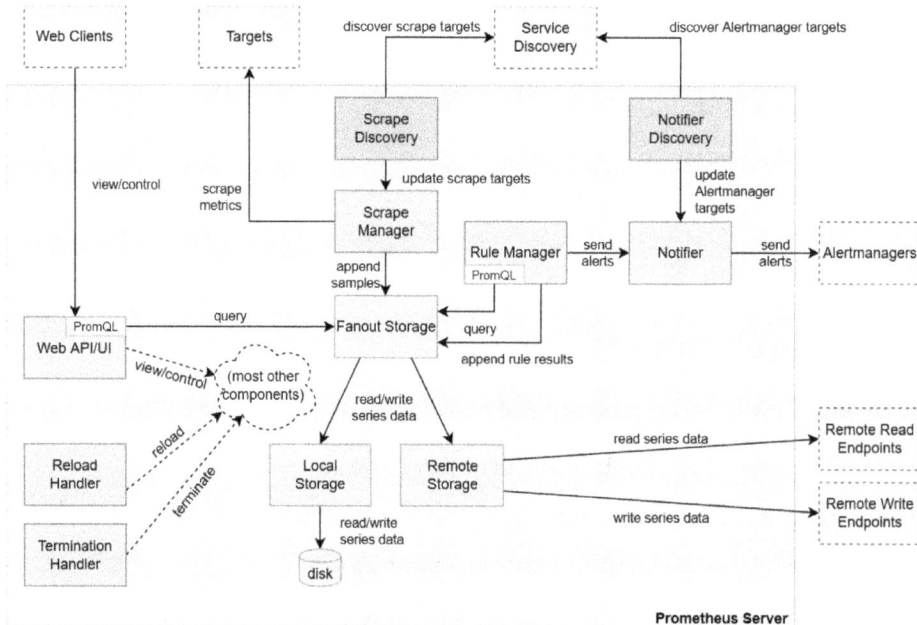

Figure 2.1 – Prometheus's internal architecture (source: https://github.com/prometheus/prometheus/blob/86a7064/documentation/internal_architecture.md)

Let's quickly go through the main components.

TSDB

The defining feature of Prometheus is its TSDB. A TSDB is a special kind of database that is optimized for storing data points chronologically. The rest of Prometheus revolves around this TSDB – the scrape manager puts things in it, the rule manager evaluates its contents, and the web UI allows for it to be queried ad hoc style. We'll dive more into the data model of the TSDB in the next chapter, so for now, all you need to know is that it's where Prometheus stores all of its data.

Scrape manager

The scrape manager is part of Prometheus that handles pulling metrics from applications and exporters both by actually performing the scraping and maintaining an internal list of what things should be scraped by Prometheus. A "scrape" is just a Prometheus term for collecting metrics. It is akin to the concept of web scraping, similar to what is done by search engines, in which data is extracted from a website in an automated way. Since Prometheus pulls metrics over HTTP (as opposed to some special protocol format or gRPC), the core concept is the same.

Rule manager

The rule manager is the piece of Prometheus that handles evaluating alerting and recording rules in Prometheus. We'll dig into recording and alerting rules later in this book, but you can think of recording rules as queries that produce new metrics (generally via aggregation) on a defined interval. On the other hand, alerting rules are queries that have a threshold attached to them to ensure that they only return data if something is wrong and an alert needs to be sent.

The rule manager handles evaluating **rule groups** comprised of alerting and/or recording rules on regular intervals (specified per rule group and/or via the global default). Additionally, it is responsible for passing along alerts that need to be sent to Alertmanager.

Web UI/API

The web UI is the portion of Prometheus that you can access via your browser. It provides helpful information for running Prometheus, such as its running configuration (including command-line flag values), TSDB information, and a simple query frontend that allows you to run ad hoc **Prometheus Query Language (PromQL)** queries to investigate your available metrics.

The Prometheus web UI is powered by the Prometheus REST API, which provides a standardized way for other programs – such as Grafana – to interact with Prometheus by querying metrics, displaying monitoring targets, and more.

Alertmanager

Alertmanager is an often overlooked piece of a Prometheus stack since those unfamiliar with Prometheus presume that Prometheus itself is responsible for sending alerts to Slack, PagerDuty, and other alerting destinations. In reality, Alertmanager handles all of that on its own through using a **routing tree-based** workflow. Prometheus itself is only responsible for evaluating alerting rules and sending alert payloads to Alertmanager, which then sends the alerts to their actual destinations. Consequently, Alertmanager is *required* if you want to enable alerting functionality in Prometheus.

Node Exporter

Node Exporter is the most prevalent Prometheus exporter. It covers (almost) anything you could need in terms of system metrics. For example, it covers metrics on CPU, RAM, disk, and network I/O in addition to a multitude of other metrics. Effectively, almost everything in the /proc pseudo-filesystem on Linux is exposed via Node Exporter (or can be via a PR to the project).

> **Node Exporter on Windows?**
>
> Node Exporter is only designed to be deployed on Linux, BSD, and Mac-based operating systems. For a Windows equivalent, please use Windows Exporter (https://github.com/prometheus-community/windows_exporter).

Naturally, this can be a significant amount of data, so many of Node Exporter's collectors are disabled by default and require a command-line flag to be enabled. The most common collectors you will care about (CPU, disk, and so on) are enabled by default. In my experience, the only additional flags I need to set for Node Exporter are to enable the systemd collector and specify a directory for the textfile collector. For the most up-to-date list of default enabled/disabled collectors, reference the README file in this project's GitHub repository or – if you have Node Exporter installed – the output of node_exporter --help.

Textfile collector

The textfile collector in Node Exporter is a unique collector that is worth drawing special attention to. A common question surrounding Prometheus is how to monitor short-lived processes and batch jobs. For this, two options exist: the textfile collector and Prometheus Pushgateway (https://github.com/prometheus/pushgateway). Of these, I generally recommend the textfile collector.

The textfile collector is enabled by default in Node Exporter but requires specifying a directory via the --collector.textfile.directory flag. Once a directory has been specified, Node Exporter will automatically read any files with a .prom file extension in that directory and include the metrics within them, alongside the typical metrics it exposes from /proc (presuming they are properly formatted).

> **Formatting text files for Prometheus**
>
> For the latest details on how to properly format text files for Prometheus' consumption, reference the Prometheus documentation located at `https://prometheus.io/docs/instrumenting/exposition_formats/#text-based-format`.

Using this, you can enable short-lived processes such as cron jobs to write out Prometheus metrics and monitor them. For some examples of writing out metrics for the textfile collector, reference the Prometheus Community project's repository, which includes helpful scripts: `https://github.com/prometheus-community/node-exporter-textfile-collector-scripts`.

Grafana

While the Prometheus web UI is great for running ad hoc queries, it has no way of storing commonly used queries, substituting variables to reuse queries, or presenting data without requiring prior knowledge of PromQL. Luckily for you, the Grafana project exists and does all of these things!

First and foremost, Grafana is a data visualization platform. It *does* include additional features such as alerting, but we won't be covering that in this book. Prometheus is the most commonly used data source type for Grafana and, as such, has great built-in functionality for visualizing time series data.

Using Grafana, we can build dashboards that include dozens of panels representing different PromQL queries and present them all on a nice, pretty web page. That's exactly what we're going to do, so let's get started.

Provisioning Kubernetes

Let's dive into setting Prometheus up! But first, we need to set up some infrastructure so that we can run Prometheus.

Throughout this book, I'll be running Prometheus on Linode using LKE. However, the same principles can be transferred to other Kubernetes environments (AKS, EKS, Minikube, and so on). Additionally, since we'll be using the Prometheus Operator – which abstracts away some pieces of Prometheus – I'll mention how to do things when running the Prometheus stack directly on Linux too.

To begin, you can register for a Linode account and get $100/60-day credit by going to `https://www.linode.com/mastering-prometheus`.

Configuring the linode-cli

With your Linode account created, we'll install the Linode CLI to handle creating our Kubernetes infrastructure via the command line.

Presuming you have Python 3 installed, the Linode CLI can be installed via `pip`:

```
$ pip install linode-cli
```

At this point, you can connect it to your account by running the following command and following the prompts in your web browser:

```
$ linode-cli configure
```

Creating a Kubernetes cluster

Now that our CLI tool has been configured, we can use it to create a Kubernetes cluster.

We'll create a cluster of three worker nodes with 4 GB of memory, two CPU cores, and 80 GB of disk space. I'll be making mine in the us-southeast region, but feel free to choose one closer to you, such as ap-west for India or eu-west for the UK:

```
$ linode-cli lke cluster-create \
    --label mastering-prometheus \
    --region us-southeast \
    --k8s_version 1.26 \
    --node_pools.type g6-standard-2 \
    --node_pools.count 3
```

It'll take a few minutes for the cluster to provision and your Linodes to show as running. Now, all we need is our Kubeconfig and we'll be ready to start deploying things:

1. To do so, find your cluster ID:

    ```
    $ linode-cli lke clusters-list
    ```

2. Save the Kubeconfig:

    ```
    $ mkdir -p ~/.kube
    $ linode-cli lke kubeconfig-view <cluster_id> --text | tail -1 |
    base64 -d > ~/.kube/config-mastering-prometheus
    ```

3. Set the appropriate permissions on it:

    ```
    $ chmod 0600 ~/.kube/config-mastering-prometheus
    ```

4. Finally, export the environment variable so that kubectl knows how to use it:

    ```
    $ export KUBECONFIG=~/.kube/config-mastering-prometheus
    ```

You can confirm that your settings are correct by trying to retrieve a list of pods (there should be none):

```
$ kubectl get pods
```

With our Kubernetes cluster set up, we can begin to set up the Prometheus Operator, which will deploy and manage Prometheus in our cluster.

Deploying Prometheus

Our Prometheus deployment will be done using a Helm chart that installs **kube-prometheus**, which installs and builds upon the Prometheus Operator by providing a convenient set of defaults. We'll begin by briefly going over what you're going to be deploying so that you understand it instead of blindly applying it because someone in a book told you to.

Prometheus Operator overview

The Prometheus Operator is a project that was originally started by Prometheus maintainer Frederic Branczyk to make it simpler to deploy and manage multiple Prometheus instances in a Kubernetes cluster. Consequently, it abstracts away some pieces of Prometheus behind various Kubernetes **Custom Resource Definitions (CRDs)**.

The main CRDs to be aware of with the Prometheus Operator are `Prometheus`, `ServiceMonitor`, `PrometheusRule`, and `Alertmanager`.

`Prometheus` and `Alertmanager` are straightforward, but `PrometheusRule` and `ServiceMonitor` are key abstractions. `PrometheusRule` abstracts Prometheus rule definitions so that they get automatically picked up by Prometheus without us needing to update Prometheus's configuration file. `ServiceMonitor` functions similarly, but with Prometheus scrape jobs instead of rules. It enables us to configure scrape jobs that correspond to Kubernetes `Service` objects that should be monitored.

> **Note**
> There is also a `PodMonitor` CRD for – you guessed it – monitoring Pods.

In addition to these CRDs, the Prometheus Operator provides other useful Kubernetes resources, such as admission webhook controllers for validating `PrometheusRule` (and other) objects for syntax and the operator itself – the controller for the CRDs managed by the Prometheus Operator.

kube-prometheus

kube-prometheus is another project run by the Prometheus Operator team but provides an opinionated way to get started with running the Prometheus Operator.

In addition to installing the operator and its associated CRDs, kube-prometheus sets up a full Kubernetes monitoring stack using it by deploying Grafana, Prometheus, and Alertmanager, and providing a base set of Prometheus rules and alerts, scrape jobs, and Grafana dashboards.

It is written using a language called **jsonnet**, which is a language that focuses on templating JSON data (which can then be output to YAML for Kubernetes thanks to YAML being a superset of JSON). This can make it difficult to customize the kube-prometheus-generated YAML manifests without knowing jsonnet yourself (don't worry – we'll learn how to use it later in this book), which is where the kube-prometheus Helm chart we'll be using comes in.

Deploying kube-prometheus

Building off the work of the kube-prometheus project, the Prometheus community maintains a Helm chart for installing it. Since the majority of Kubernetes users are far more familiar with Helm than jsonnet, we'll be using that to deploy the kube-prometheus stack.

Configuration

To install the chart, we'll need to add the Helm repository for it:

```
$ helm repo add prometheus-community https://prometheus-community.
github.io/helm-charts
$ helm repo update
```

Next, we need to configure the chart how we want it by overriding the default values in a values file as needed. You can certainly apply it with the defaults, but the changes I will make are available in this book's GitHub repository.

I changed a setting in the Prometheus Operator so that it discovers `ServiceMonitor` objects in all namespaces, and changed some Grafana settings such as the default user and password, as well as the default time zone for dashboards.

To see all the available configuration values, you can run the following command:

```
$ helm show values prometheus-community/kube-prometheus-stack
```

Deployment

With our values set, we can trigger the installation:

```
$ helm install \
    --namespace prometheus \
    --create-namespace \
    --values mastering-prometheus/ch2/values.yaml \
    --version 47.0.0 \
    mastering-prometheus \
    prometheus-community/kube-prometheus-stack
```

If everything went well, you should now have Alertmanager, Prometheus, Grafana, and a few other things running on your cluster! You can confirm this by switching to the `prometheus` namespace and listing your pods:

```
$ kubectl config set-context --current --namespace Prometheus
$ kubectl get pods
```

Validation

Now that we've got kube-prometheus running in Kubernetes, we can see what it's doing. The easiest way to do so is by opening the Grafana instance we just created. Proxy connections to it with the following command and log into Grafana in your browser at `http://localhost:3000`:

```
$ kubectl port-forward svc/mastering-prometheus-grafana 3000:80
```

You should see a landing page that looks like this:

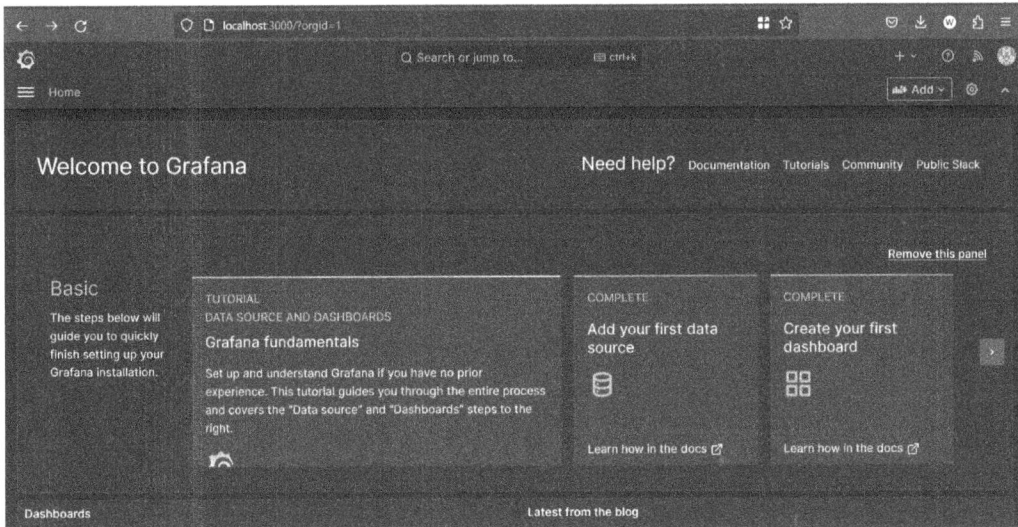

Figure 2.2 – Fresh Grafana installation

Grafana is pre-populated with helpful dashboards for Kubernetes and Node Exporter data. A bonus application that is included with kube-prometheus is the **kube-state-metrics** service, which provides a tremendous amount of data (2,300+ time series in the cluster we just created) about your Kubernetes cluster in Prometheus format.

Between the Node Exporter metrics, cAdvisor metrics exposed by kubelet, and kube-state-metrics, you've practically covered all of your infrastructure metrics inside of the cluster with that simple Helm deploy.

As an example of the data available, open the **Kubernetes / Compute Resources / Cluster** dashboard and click around to explore your data using all of the built-in drill-down links:

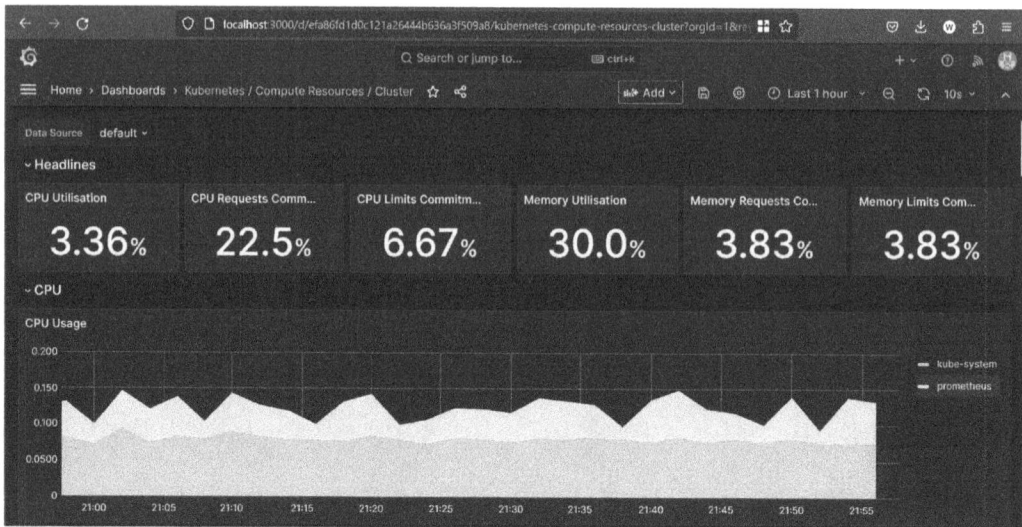

Figure 2.3 – The Kubernetes / Compute Resources / Cluster dashboard

Now that we've explored all of the glorious data we have been blessed with, let's get ready to dig into the next chapter.

Summary

In this chapter, we went over the key components of a basic Prometheus stack: Prometheus, Alertmanager, Node Exporter, and Grafana. We looked at Prometheus's internal components to understand what it does and what it delegates to other services, such as Alertmanager. Finally, we went hands-on and built a Kubernetes cluster and Prometheus stack from the ground up in record time.

Going forward, we'll build upon this architecture to see how Prometheus scales and how to extend it. But first, we're going to dive into some of the nitty-gritty of Prometheus by focusing on its data model and query language so that we can get the most out of the data we're collecting from Prometheus.

Further reading

To learn more about the topics that were covered in this chapter, take a look at the following resource:

- *The prometheus-operator project*: `https://github.com/prometheus-operator/prometheus-operator`

3
The Prometheus Data Model and PromQL

With Prometheus deployed, let's explore a little more about what it's doing under the hood. You're likely already somewhat familiar with Prometheus, but you may not know the intricacies of how it works. It's something I'm still learning every day and I dare say that Prometheus's code base has grown to the size where there may not be a single person or maintainer who understands the ins and outs of how every piece of it works. Nevertheless, there's still plenty to be learned; the more you understand a system, the better you can run and maintain it.

The core of the Prometheus system is its TSDB. Most of the work that Prometheus does is centered around either getting data into the TSDB or reading data out of it. But TSDBs are complicated topics that are ever-changing and evolving. Even Prometheus underwent a major TSDB rewrite when it went from version 1 to version 2. So, we'll keep our inquiry focused on Prometheus's TSDB, its data model, and the tradeoffs it makes.

Once we understand more of how the TSDB works, we'll look at how to pull data out of it using PromQL. You'll learn some neat tips and tricks for crafting some expert-level queries that you can use for your alerts and dashboards.

In this chapter, we will cover the following topics:

- Prometheus's data model
- Prometheus's TSDB
- PromQL basics

Let's get started!

Technical requirements

Presuming you installed Prometheus using the guide in the previous chapter, you will only need the following program to be installed and available in your PATH so that you can connect to Prometheus:

- **kubectl**: `https://kubernetes.io/docs/tasks/tools/#kubectl`

Please refer to this book's GitHub repository for additional resources: `https://github.com/PacktPublishing/Mastering-Prometheus`

Prometheus's data model

Prometheus's data model is – in my opinion – straightforward. It has three main data types that are all built on each other: *metrics*, *time series*, and *samples*.

Metrics

Metrics are the fundamental Prometheus data type. After all, Prometheus is a *metrics*-focused observability tool.

Every metric has – at a minimum – a name that identifies it. This name can contain letters, digits, underscores, and/or colons. To be valid, it must match the `[a-zA-Z_:][a-zA-Z0-9_:]*` regex.

Additionally, metrics may also include a HELP text and a TYPE field. These are optional but highly recommended to improve usability.

A full metric that's been exposed to Prometheus may look like this:

```
# HELP mastering_prometheus_readers_total Number of readers of this
book
# TYPE mastering_prometheus_readers_total counter
mastering_prometheus_readers_total 123467890
```

The HELP line just provides some arbitrary information on what the metric is intended to represent, which can be helpful to consumers of your Prometheus data when determining if a particular metric is relevant to them.

The TYPE line indicates what type of metric your metric is. The type doesn't affect how the data is stored in Prometheus's TSDB at all – it's all the same under the hood. However, all of the Prometheus client libraries for instrumenting applications use the same common metric types to help organize metrics into different classifications. Additionally, it makes it easier to reason about which PromQL functions to use with a metric. Certain PromQL functions, such as `rate` or `histogram_quantile`, are designed to only work with specific metric types and will return incorrect or unexpected results if used with metric types they were not designed for. At the time of writing, there are four core metric types in Prometheus: counters, gauges, histograms, and summaries.

Counters

Counters are the simplest Prometheus metric type. They start at zero and can only go up during a process's lifetime. They can easily be identified by the semantic convention of ending their metric name in a `_total` suffix. Possible usages include error counters or counting the number of requests received by a web server.

Gauges

Gauges are similar to counters in their simplicity, but they allow for values that can go up *or down*. Possible uses include tracking active connections to a web server or the memory usage of a system. Additionally, they are commonly used to track Boolean values such as in the built-in `up` metric for Prometheus scrape jobs, where a value of `0` indicates a scrape job is down and a value of `1` means it is up. Finally, gauges are also frequently used for information style metrics, which is a pattern of defining metrics with a `_info` suffix, where the value is always set to `1` and the labels provide additional information about an application or server. An example of this is Prometheus's `prometheus_build_info` metric, which – among other things – includes labels such as `version` to report the running Prometheus version.

Histograms

Histograms and summaries are where things get more complicated. A histogram samples observations of values and counts them in configurable buckets. In a way, a histogram is a collection of multiple counters.

Histograms get a label called `le` that breaks up buckets by thresholds. A common usage is to track request latencies. You may have a histogram named `http_request_duration_seconds_bucket` with five buckets: `[ 0.1, 0.2, 0.5, 1, 5 ]`. For each observation, each bucket's counter is incremented by 1 if the observation is less than or equal to (hence `le`) the bucket threshold. For example, a request with a latency of 0.6 seconds would increment the `le="1"` and `le="5"` buckets. Alternatively, a request with a latency of 0.05 seconds would increment all of the buckets.

Additionally, histograms get a special `le` bucket called `le="+Inf"`, where `Inf` is short for infinity. All observations increment this bucket's value, so it's akin to a total counter for the metric. There's also a metric for `<histogramName>_count` that is equal to `<histogramName>_bucket{le="+Inf"}`, and a `<histogramName>_sum` metric that is the sum of all observed values.

So, what makes histograms so useful? Are they worth the added complexity? The primary benefit I've found of histograms is the ability to calculate histogram quantiles, which can be useful for measuring things such as tail latency and SLOs. Since observations are bucketed, it enables neat querying tricks to answer questions such as, "What is my 99th percentile latency on this service?"

Summaries

Summaries are similar to histograms in that they also are derived from observed values. However, they do some precomputation to expose special quantile labels instead of `le` labels. Therefore, when defining a summary, those quantiles need to be specified in advance rather than bucket thresholds. Those quantiles are specified as values from 0 to 1 and represent percentages. For example, a metric may have quantiles of [0.01, 0.05, 0.5, 0.9, 0.99]. The values for each of these quantiles are recomputed with each observation.

In practice, summaries are the least used metric type that I see and are generally reserved for very specific use cases.

Time series

All right, enough about metric types – back to those different core Prometheus data types we were talking about! Time series are the next Prometheus data type we'll discuss since they are built on top of metrics. One metric can have dozens of time series. But how? Through the usage of **labels**!

Labels are key/value pairs that add additional detail to a metric. For example, a metric may be collected from multiple servers. To distinguish them, an instance label is added to the series relating to the metric to reflect where that series came from. It may be better to just visualize it:

Metric	node_cpu_seconds_total
Time Series	node_cpu_seconds_total{cpu="0",instance="serverA"}

The way I think of it is that a metric is just the name – nothing within the curly braces (although, as we'll see later, the name itself is stored as a hidden label). If you're familiar with object-oriented programming, you can think of a metric as a base class and a time series as a *unique* instance of that class. That uniqueness is also important.

All series that are exposed to Prometheus in a scrape job must be unique, so series with the same metric name should be grouped under a single HELP/TYPE definition and have unique labels. If they are not unique, the Prometheus scrape will fail. For example, the following would be invalid since the time series is duplicated without unique labels:

```
# HELP mastering_prometheus_readers_total Number of readers of this
book
# TYPE mastering_prometheus_readers_total counter
mastering_prometheus_readers_total 123467890
mastering_prometheus_readers_total 789
```

However, this *would* be valid since the individual series for the metric is unique:

```
# HELP mastering_prometheus_readers_total Number of readers of this
book
# TYPE mastering_prometheus_readers_total counter
mastering_prometheus_readers_total{stage="published"} 123467890
mastering_prometheus_readers_total{stage="draft"} 789
```

Time series are tracked over time through a series of samples, which brings us to our final data type: samples.

Samples

Samples are built on top of time series, just like how time series are built on top of metrics. A metric isn't much use without a time series that tracks it, and a time series isn't much use without samples that represent its values over time.

Each time that Prometheus scrapes a target, it collects samples for all of the time series returned by that scrape target. A sample is extremely simple: it's just a tuple of a timestamp and a value. That's it!

Commonly, samples are denoted using the `(t0, v0), (t1, v1), . . .` format, where `tX` is a timestamp and `vX` is a value.

Values are represented in a `float64` type in Go (this means it has a decimal point) and timestamps are represented using the number of milliseconds since the Unix epoch (January 1, 1970).

Now that we understand Prometheus's data model, let's see how it all comes together inside its TSDB.

Prometheus' TSDB

Prometheus's TSDB is at the core of all that Prometheus does. It's been in its third iteration since 2017 and continues to be regularly updated with new optimizations and features. However, its core functionality and format have largely remained the same.

Understanding how the TSDB works will enable you to understand Prometheus's internals in a new way that can inform how you maintain and use Prometheus. The core parts of the TSDB that we'll look at are its head block, **write-ahead log (WAL)**, and data format (blocks, chunks, and indices), as well as how compaction works.

Head block

The head block in Prometheus is the first stop for samples being scraped and stored in Prometheus' TSDB. It is an in-memory block (more on the specifics of a block later) containing various chunks for each of the different time series you're collecting. This block can contain millions of chunks, depending on the number of scrape targets and time series you have.

When a sample is scraped by Prometheus, it is appended to the appropriate *active* chunk for the time series the sample pertains to. Remember, a time series is distinct from a metric.

So, each time series gets a chunk that samples are appended to in the head block as they come in. Each chunk for a series in the head block can grow to 120 samples before it is automatically flushed to the active chunk's **segment file** in the `chunks_head` directory on the disk. For a scrape interval of 30 seconds, this would be 1 hour of data in a chunk before it's flushed to a file.

> **Head chunk segment files**
>
> Rather than maintain a file for each time series, Prometheus puts multiple chunks inside of segment files for storage. Time series need not be related to each other in any way to end up in the same segment file. Which segment file a chunk goes into is determined by which segment file was being actively appended to when the chunk was flushed. Each head chunk segment file can grow to a maximum of 128 MiB before a new file is started.

As shown in the following figure, where my example Prometheus setup is using a 30-second scrape interval, the number of chunks is roughly 2x the number of series, which makes sense since in 2 hours (the default minimum size of a persistent block), I would have two full chunks for each series. If your scrape interval was 15 seconds, you would have several chunks equal to about 4x the number of series, and so on:

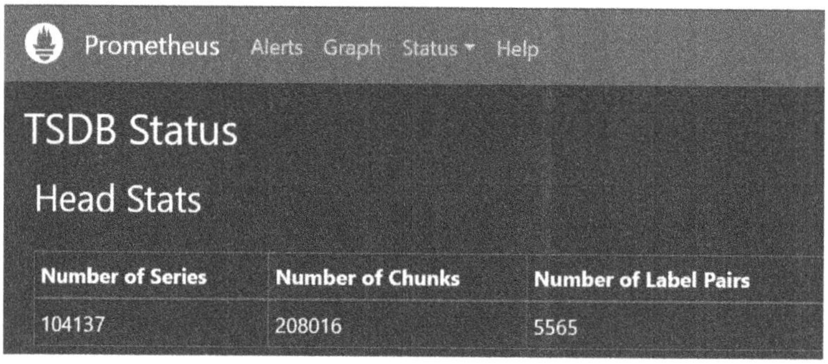

Figure 3.1 – Prometheus head block statistics

Once flushed to the chunks file on disk, the chunk is then **memory-mapped (mmap'd)** to the head block instead of continuing to hold the whole thing in memory. The head block stores the minimum and maximum time for the chunk and a reference that contains information about which file the chunk was written to and the offset within the file at which the chunk starts.

Finally, when the data in the head block spans 1.5x the minimum block range (which is 2 hours by default and assumed to be used in this example), the first 2 hours in the head block are compacted and written to a **persistent block** on disk. At this time, the WAL is also truncated, and a checkpoint is created. Next, we'll look at what the WAL is doing during all this.

WAL

When version 2 of Prometheus launched, the WAL was one of the flagship features that improved on version 1. In version 1 of Prometheus, you could potentially lose metric data if Prometheus crashed between intermittent checkpoints that would be performed. With the WAL, Prometheus users no longer need to worry about data loss from application-level crashes.

We've already discussed how when a sample comes into the TSDB head block, it gets appended to a chunk in the head block. Until that chunk hits its sample limit (or maximum age), it's just in memory. This means that if Prometheus restarts or our server crashes, we lose that data! Well, we would if it weren't for the WAL.

As it turns out, before a sample is even added to an in-memory head chunk, it is first appended to the WAL. This provides an added layer of durability (the D in ACID, for the database nerds out there) for writes.

The Prometheus WAL is only used to provide this durability guarantee, so it is only ever read from to restore the in-memory state of the Prometheus server during startup.

Similar to the storage of head chunk files on disk, the WAL is also broken up and stored in 128 MiB files, like so:

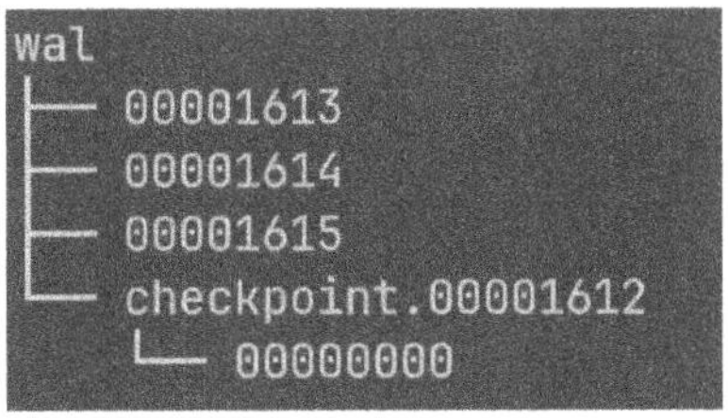

```
wal
├── 00001613
├── 00001614
├── 00001615
└── checkpoint.00001612
    └── 00000000
```

Figure 3.2 – WAL directory structure

The files, also like chunk files, have numerical names that increase sequentially. However, you'll notice that I don't have files named 00000001, 00000002, and so on in the preceding screenshot. Where did they go?

Once a persistent block is written out to the disk, the portion of the WAL that pertained to that time range is no longer needed since all the data has been flushed out of memory. Consequently, WAL truncation occurs immediately after the head block is truncated. However, the method of truncation differs.

Whereas head block truncation can safely delete over a time range, the same is not true of the WAL. It's non-trivial to determine the time range of a given WAL segment file, so instead, Prometheus just deletes the first two-thirds of the segment files. In my example, that would be 00001613 and 00001614. Yet that may not be entirely safe because there may still be samples in those files that haven't been truncated from the head block yet. To solve this issue, a WAL "checkpoint" is created from the segments that are going to be deleted.

Checkpoints

A WAL checkpoint is described as a "filtered" WAL, in which the to-be-deleted segments are iterated over and the following operations are performed:

1. Drop series records if the series is not in the head any longer.

2. Drop samples from before the timestamp at which truncation is occurring.

3. Drop tombstone records from before the timestamp at which truncation is occurring.

4. Keep everything else the same (and in the same order).

Checkpoints are integral to the process of speedily restoring a Prometheus instance to an active state by "replaying" the WAL when it starts up.

Replay

When a Prometheus server is restarted, the WAL is critical to getting it back into the state it was in before it was shut down without losing data. Unfortunately, with a lot of series and samples, replaying the WAL can take a while. In older versions of Prometheus, it was not uncommon for me to see WAL replays take more than 30 minutes when a Prometheus instance was restarted. However, a significant amount of optimization has been made to the replay procedure since and those same servers complete WAL replays in around 5 minutes.

What is Prometheus doing that's taking so long? Well, multiple pieces play into replay on server startup. The order of operations looks roughly like this:

1. Load mmap'd chunks from disk.

2. Replay checkpoint contents.

3. Replay all other WAL segments from after the checkpoint.

There is a caveat to this order of operations if you have the optional Prometheus feature flag enabled to take a snapshot of in-memory data (plus series information) on shutdown (`--enable-feature=memory-snapshot-on-shutdown`). With this feature enabled, Prometheus takes a snapshot of the chunk data it has in memory when it's told to shut down, which enables a quicker restoration of the in-memory state when the process starts again. This is because it will generally skip the need to read from the WAL at all during startup. In this case, the process would look like this:

1. Load mmap'd chunks from disk.

2. Iterate through the chunk snapshot to restore the series.

3. Replay any WAL data after the snapshot.

4. Generally, no WAL replay will be done since the maximum time of the WAL should be less than/equal to the maximum time of the snapshot.

But what about historical Prometheus data? The data that isn't in the head block or the WAL? Let's take a look at what persistent data looks like on Prometheus by exploring blocks and chunks on disk.

Blocks and chunks

We alluded to blocks and chunks when we discussed the head block and WAL, but let's expand on them further and focus on how they work on disk.

The blocks and chunks we'll be talking about in this section are the *persistent blocks* and *persistent chunks* that are written to disk when the head block is truncated. Blocks and chunks are **immutable**, meaning that they cannot be modified once they are written. To make any alterations to a persistent block, you would have to rewrite the entire block.

Each block is comprised of four pieces:

- An index file

- A directory of chunk segment files

- A metadata file

- A tombstones file

Block names themselves are **Universally Unique Lexicographically Sortable Identifiers (ULIDs)**.

On disk, it looks something like this:

```
01H423RRB80ER7EDB91WPY1H39/
├── chunks
│   └── 000001
├── index
├── meta.json
└── tombstones
```

Figure 3.3 – Block directory structure

Combined, these four parts allow each block to stand on its own – independent of any other blocks. Essentially, each block is a fully independent database.

Tombstones

Tombstones are probably the least interesting thing in a block, but they are still useful. As mentioned previously, blocks are immutable. However, there may still be cases in which you want to delete a series from Prometheus. If we had to rewrite every block where a series is present to delete it, it would be extremely computationally expensive. So, instead, Prometheus's TSDB uses tombstone markers to indicate that a series should be ignored during reads.

Tombstones are not all-or-nothing, though. In addition to what series to ignore, they also include a minimum time and a maximum time to indicate for which time range that series should be ignored.

> **Fun fact**
>
> The `tombstone` file is the only file in a persistent block that is modified after the block has been created and written to disk.

Metadata

The metadata for a block is stored in the `meta.json` file and it is exactly what you would expect. It contains basic information about a block, such as its unique ID, its minimum and maximum times, and some stats on the number of samples, series, and chunks it contains.

A fresh block's `meta.json` file looks something like this:

```json
{
    "ulid": "01H4SQ2CFN7RG9MT4C1ERS83WG",
    "minTime": 1688774400135,
    "maxTime": 1688781600000,
    "stats": {
        "numSamples": 24998218,
        "numSeries": 104180,
        "numChunks": 208270
    },
    "compaction": {
        "level": 1,
        "sources": [
            "01H4SQ2CFN7RG9MT4C1ERS83WG"
        ]
    },
    "version": 1
}
```

Figure 3.4 – A fresh TSDB block's meta.json file

However, as we'll explore later, blocks are also compacted regularly by Prometheus. Since these compacted blocks are a combination of multiple blocks, their `meta.json` file looks different:

```
{
        "ulid": "01H423RRB80ER7EDB91WPY1H39",
        "minTime": 1687910400322,
        "maxTime": 1687975200000,
        "stats": {
                "numSamples": 224317360,
                "numSeries": 103912,
                "numChunks": 1868915
        },
        "compaction": {
                "level": 3,
                "sources": [
                        "01H3ZZ34HM66FSGF6SQZST2E6D",
                        "01H405YVS0RE9N3NZBP4XM3422",
                        "01H40CTK1SWSZDR052RJZPR778",
                        "01H40KPAS77AK9N96JC2HYZDHM",
                        "01H40TJ1JWGSW2MZRDV4KK0VRZ",
                        "01H411DRSFWJPJEDZ2C3K3BC79",
                        "01H4189G1F3RNNEM0ESQSCQGHZ",
                        "01H41F579ZDMKQYW9338T8J62A",
                        "01H41P0Z3AYYEHYH303CSQ9V83"
                ],
                "parents": [
                        {
                                "ulid": "01H40TJ9VFSEMKQRVJFB0SF5XD",
                                "minTime": 1687910400322,
                                "maxTime": 1687932000000
                        },
                        {
                                "ulid": "01H41F5J0XD5MAMKDK3HFAP22S",
                                "minTime": 1687932000316,
                                "maxTime": 1687953600000
                        },
                        {
                                "ulid": "01H423RJQBYGGQ9AZV3XV0XG52",
                                "minTime": 1687953600276,
                                "maxTime": 1687975200000
                        }
                ]
        },
        "version": 1
}
```

Figure 3.5 – A level three compacted block's meta.json file

If a block is essentially a data structure, then chunks are the data within it. Now that we understand the structure of a block, let's look at the chunks it stores.

Chunks

Chunks are what a block is all about – they are stored in the `chunks/` directory of the block in **segment files**, which contain a bunch (potentially millions) of individual chunks within them. Doing the math on a production Prometheus server I have, each chunk segment file in a block has almost four million chunks within it!

Each chunk segment has a maximum size of 512 MiB by default, but this can be configured via the `--storage.tsdb.max-block-chunk-segment-size` flag (a minimum of 1 MiB). This is typically used for smaller Prometheus installations where the 512 MiB default results in unnecessary overhead when pre-allocating space for chunk segment files.

> **Remember**
>
> The files in the `chunks` directory are *not* chunks themselves. Each file is comprised of numerous unrelated chunks.

The format of the chunks within the chunk segments file is very similar to the format of chunks in the head block but with a little less data stored directly in the chunk since that data is shifted to the block's index. Namely, head block chunks store information about the minimum and maximum time for samples in the chunk, along with a reference to the time series they pertain to, but persistent chunks don't need that information.

An individual chunk within a chunk segment file is just a collection of timestamps and values since each sample from a scrape is just a value with a timestamp attached. Well, there's a little more nuance to this.

To increase resource efficiency, some wizardry is done to the chunk data so that it only stores the direct value of the first timestamp (`t0`) and first value (`v0`). Subsequent timestamps are set to the delta of the prior timestamp, so starting at `t2`, all timestamps are the delta of a delta. Similarly, subsequent values are compared to their prior value using a bitwise `XOR` operator. This just means that the difference between the samples is stored, and if they're the same, then 0 is stored.

This usually comes into play with timestamp deltas. If all the deltas are 0 (as they ought to be since scrapes should be happening at consistent intervals), then that chunk's data compresses much more efficiently. Otherwise, you can end up with blocks that take up significantly more disk space than you might expect.

Index

The index is like the table of contents for a block. Well, the index itself has a table of contents… so maybe not. Let's just stick with calling it an index. And to be technically correct, it's an *inverted* index.

Without the index, a block would be unusable as any client reading from the TSDB would have no idea what series are in the block or where to find the sample values pertaining to them.

The index is comprised of multiple different sections:

- A **Series** section
- A **Postings** section
- A **table of contents** (TOC)

On disk, the format looks like this:

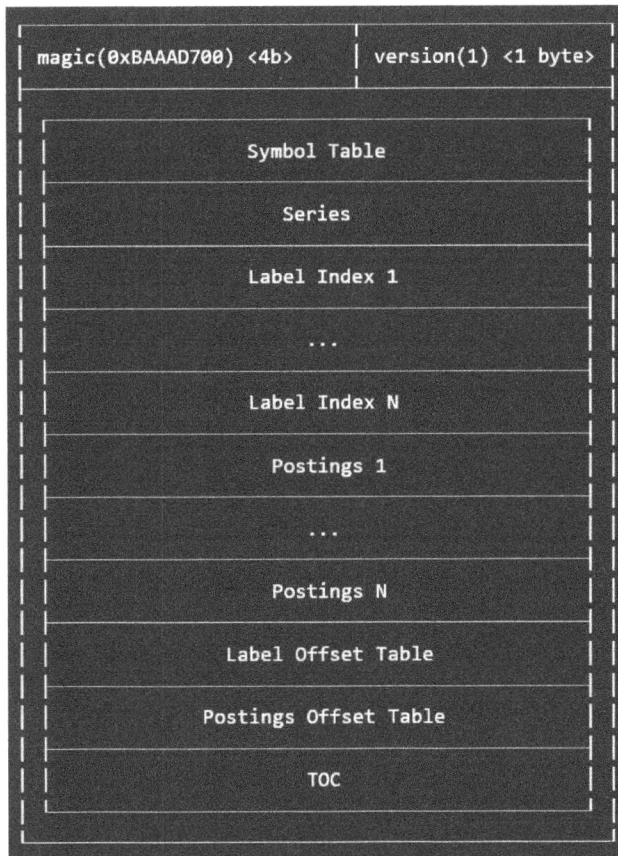

```
┌─────────────────────────────────┬──────────────────────────┐
│ magic(0xBAAAD700) <4b>          │ version(1) <1 byte>      │
├─────────────────────────────────┴──────────────────────────┤
│  ┌───────────────────────────────────────────────────────┐ │
│  │                  Symbol Table                         │ │
│  ├───────────────────────────────────────────────────────┤ │
│  │                     Series                            │ │
│  ├───────────────────────────────────────────────────────┤ │
│  │                  Label Index 1                        │ │
│  ├───────────────────────────────────────────────────────┤ │
│  │                       ...                             │ │
│  ├───────────────────────────────────────────────────────┤ │
│  │                  Label Index N                        │ │
│  ├───────────────────────────────────────────────────────┤ │
│  │                    Postings 1                         │ │
│  ├───────────────────────────────────────────────────────┤ │
│  │                       ...                             │ │
│  ├───────────────────────────────────────────────────────┤ │
│  │                    Postings N                         │ │
│  ├───────────────────────────────────────────────────────┤ │
│  │                Label Offset Table                     │ │
│  ├───────────────────────────────────────────────────────┤ │
│  │               Postings Offset Table                   │ │
│  ├───────────────────────────────────────────────────────┤ │
│  │                       TOC                             │ │
│  └───────────────────────────────────────────────────────┘ │
└─────────────────────────────────────────────────────────────┘
```

Figure 3.6 – Prometheus index format (source: https://github.com/prometheus/prometheus)

There is a **Label Index** section, but that is no longer used in Prometheus and is only included for backward compatibility.

TOC

The TOC of the index is contained at the end of the index file as opposed to the beginning. It is an entry point to the index and provides the byte offset within the file that other sections of the index start.

Unlike the other sections of the file, it has a fixed size since its structure is pre-determined and independent of the number of series and labels in a given Prometheus instance.

Postings

The postings index is the index that stores the relationship between labels and sets of series containing those labels. Each unique combination of labels is given a unique series ID (a posting).

> **What is a "posting"?**
>
> According to Prometheus's TSDB maintainer, Ganesh Vernekar, the term "posting" likely comes from the terminology used within the field of document indexing in which the full-text index refers to document IDs as a "posting."

The postings index is comprised of the postings offset table and a list of postings that map to series IDs. The offset table contains information for label key/value pairs:

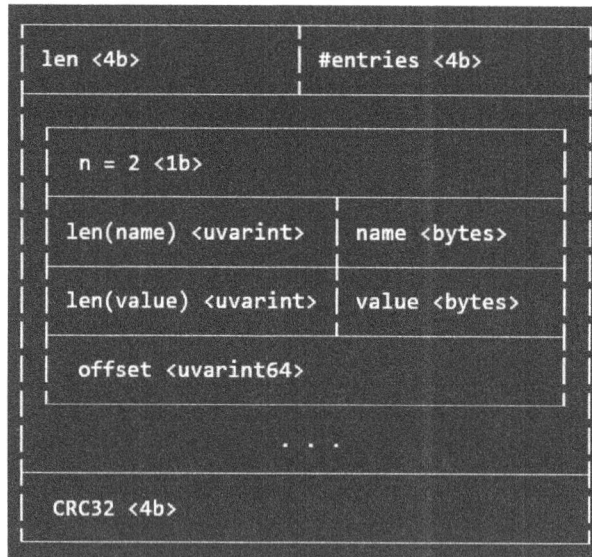

Figure 3.7 – Postings offset table entry format (source: https://github.com/prometheus/prometheus)

Each entry in the offset table contains only one label name and value. Additionally, each entry point in an offset in the `Postings` section contains a postings list of series IDs containing that label + value pairing. For example, an entry in the postings offset table may say that for a list of series matching `{job="kubelet"}`, go to offset X in the postings table.

For TSDB queries with multiple label selectors such as `{job="kubelet", env="production"}`, the postings from each label/value pair are intersected to find the postings (series IDs) that match both. For example, if `{job="kubelet"}` points to series IDs `[10, 11, 17]` and `{env="production"}` points to series IDs `[7, 9, 11, 13, 17]`, then the intersection would be `[11, 17]` and those are the series we would need to look up in the series index.

Series

After label selectors are matched via the postings index, Prometheus jumps to the aptly named series index to find the actual series data we're looking for.

The series index is comprised of entries for every unique label set. In Prometheus, metric names are internally stored as a label called `__name__`, so the metric name (for example, `node_cpu_seconds_total`) gets stored in the TSDB as an additional label in the set. Consequently, an entry may exist in the series index for a series such as `{__name__="node_cpu_seconds_total", job="node_exporter", env="production"}`.

Each entry in the series index contains information on each label, key, and value applicable to the series, the number of chunks that store data about the series, and then an entry for each of those chunks that contain information, such as the minimum and maximum time of the chunk, as well as the reference to the chunk in the `chunks/` directory.

> **Fun fact**
> The reason that the minimum/maximum time of each chunk is stored in the index is to help accelerate queries by ensuring that time is not wasted attempting to read chunks that are not within the time range of the query being executed against the TSDB.

If you're like me, you may be wondering how the chunks are referenced from the series index. Chunks within a block are stored in multiple segment files, so how does it know which file to look in and where to read it? Well, the reference (which is a `uint64` data type) is divided in half. A 64-bit unsigned integer is comprised of 8 bytes, so Prometheus uses the *upper* 4 bytes to denote which chunk segment file the chunk is stored in, and it uses the *lower* 4 bytes to specify the offset within that file that the chunk begins. Pretty nifty.

Tying it together

The index is a big topic and we just covered a lot in a relatively short span. It may be helpful to see how all its pieces fit together in the context of a query being executed against a block in the TSDB.

This is how the flow goes to retrieve query results in Prometheus for blocks matching the time range of the query:

1. Use the label matchers of the query and find series that match them using the postings index.

2. If multiple label matchers are present, use the intersection of all the matchers to find the series that contain all of the labels.

3. Using the results from the postings index, look up series in the series index to find chunks that have the data for the relevant series within the time range.

4. Retrieve samples by reading chunks using the chunk reference(s) retrieved from the series index.

5. Return sample data.

At this point, knowing how the index is created and stored, it may seem superfluous to have this same big index repeated over and over for every 2-hour TSDB block – especially if you're in an environment with low metric churn (your labels don't change often). You'd be right in thinking so! To address this, Prometheus performs compaction of blocks.

Compaction

It can be somewhat computationally expensive to run queries across multiple blocks and merge the results – especially if those blocks overlap in time and deduplication of samples has to occur. Additionally, it's not a great use of disk space to have an index for each of these small blocks that probably doesn't differ *too* much from block to block in terms of what series are present. This is where compaction comes in.

Prometheus has a concept of compaction in which multiple small adjacent blocks can be combined into one larger block. Compacted blocks are comprised of three or more smaller blocks and are compacted in durations multiplied by three starting at 2 hours by default. In other words, the time range of blocks goes from 2h → 6h → 18h → 54h, and so on.

To trigger compaction, Prometheus runs a compaction cycle every minute to see if there is any work to be done by generating a "plan" for compaction. A compaction plan checks for three conditions:

1. Overlapping blocks are present.

2. Enough blocks of smaller time ranges are present to create a block of the next longest time range – for example, >=3-hour blocks to make a 6-hour block, or >=3 6-hour blocks to make an 18-hour block.

3.   Tombstones cover more than 5% of the series in a block. This saves disk space by deleting series data, but only if a meaningful amount of disk space could be reclaimed by doing so since – as you'll recall – this requires rewriting the whole block, which is computationally expensive.

Each plan only addresses one of the three conditions specified previously, meaning that an individual compaction run will be required to address each condition and may result in multiple compactions. Additionally, they are prioritized in the order shown here. This means that overlapping blocks always take precedence in being addressed, then "normal" compaction of combining multiple small blocks into a bigger block, and finally rewriting blocks with a lot of deleted series.

Retention

Retention of blocks is closely related to compaction. Similar to compaction, it also runs every minute. However, it is concerned with deleting blocks when the TSDB surpasses at least one of two thresholds: a time-based threshold or a size-based threshold.

Prometheus has two command-line flags that control how long data is retained, and both retention settings can be applied simultaneously.

Flag	Default
`--storage.tsdb.retention.time`	15d
`--storage.tsdb.retention.size`	N/A

The time-base threshold means that blocks will be retained up to the configured length of time. In practice, this may mean that you are left with data that does not go right up to the configured retention time. This is because retention is applied based on the maximum time of the block, so the whole block does not have to be outside the retention window. In other words, a 54-hour block that starts at 2023-01-01T00:00:00 and ends at 2023-01-03T06:00:00 would be deleted if the cut-off for retention is at 2023-01-03T05:00:00, leaving you with 53 hours of data less than the configured maximum length of time.

Similarly, the size-based threshold will delete the block that causes the size threshold to be exceeded and all other blocks that are older than it. Importantly, size-based retention is applied to the oldest blocks, so you don't have to worry about Prometheus deleting blocks at random to stay within the configured retention size. Blocks are sorted based on their maximum timestamp from their meta.json file, size is calculated, and then blocks are marked for deletion using this code (as of Prometheus v2.48.0):

```
blocksSize := db.Head().Size()
for i, block := range blocks {
        blocksSize += block.Size()
        if blocksSize > db.opts.MaxBytes {
                // Add this and all following blocks for deletion.
                for _, b := range blocks[i:] {
                        deletable[b.meta.ULID] = struct{}{}
```

```
            }
            db.metrics.sizeRetentionCount.Inc()
            break
        }
}
```

> **In practice**
>
> In my Prometheus instances, I apply both time-based and size-based retention. Size-based retention is a safeguard to prevent Prometheus from filling the disk on a server and causing other issues. Ideally, it should never be hit.
>
> If it *is* hit, we configure an alert based on an increase to the `prometheus_tsdb_size_retentions_total` metric to let us know that we're losing data earlier than our configured time-based retention.

PromQL basics

With all this talk of how data gets into Prometheus's database and how it's stored there, I reckon it's about time we look at how we get that data out and make use of it. To do this, Prometheus has its own query language: PromQL.

PromQL is a vast topic, and you can craft some cool queries with it, especially concerning alerting (covered in *Chapter 5*).

> **Note**
>
> Most PromQL queries provided as examples will work in your Prometheus environment from *Chapter 2*!
>
> Feel free to tweak them and experiment with them as you go along by using `kubectl port-forward svc/mastering-prometheus-kube-prometheus 9090` and opening `http://localhost:9090` in your browser to connect to Prometheus.

Syntax overview

PromQL is pretty straightforward and does not have a litany of different data types you need to concern yourself with. The matching operators for label filters are intuitive, and there are only four possible types that PromQL expressions can be evaluated as:

- Instant vector: Explained next
- Range vector: Explained next

- Scalar: A number value

- String: A string value; at the time of writing, this is defined but unused in Prometheus

Both instant and range vectors have similar basic syntax consisting of a metric name and/or label matchers. A metric name in Prometheus is a label named __name__ under the hood, but PromQL has syntax niceties that allow you to just specify the metric name as the start of the selector. For example, the following are equivalent and return the same data:

- Explicit label matcher on metric name

```
{__name__="prometheus_build_info"}
```

- Equivalent query without an explicit label matcher syntax

```
prometheus_build_info
```

Label matchers are defined using a label key, a label matching operator, and the label value. Here are some possible label-matching operators:

- =: The label value matches the specified string

- ! =: The label value does *not* match the specified string

- =~: The label value matches the regex in the specified string

- ! ~: The label value does *not* match the regex in the specified string

For example, we could match all the pods we created in *Chapter 2* using this query:

```
kube_pod_info{pod=~".*mastering-prometheus.+"}
```

This will return all series from the kube_pod_info metric that has a pod label that matches our provided regex.

Instant versus range vectors

All Prometheus queries you will execute fall into one of two types: *instant* and *range*.

In PromQL, these types are referred to as instant vector selectors and range vector selectors. The word "query" is often used interchangeably with "vector," but we'll use vector for correctness and to help us distinguish between Prometheus's /query and /query_range API endpoints later.

Instant vectors are selectors that select data at this instant – using the most recently available data for the time series you're selecting. Consequently, they only return a single value per time series. Additionally, in the Prometheus UI, the timestamp will not be displayed alongside the value in the default **Table** pane:

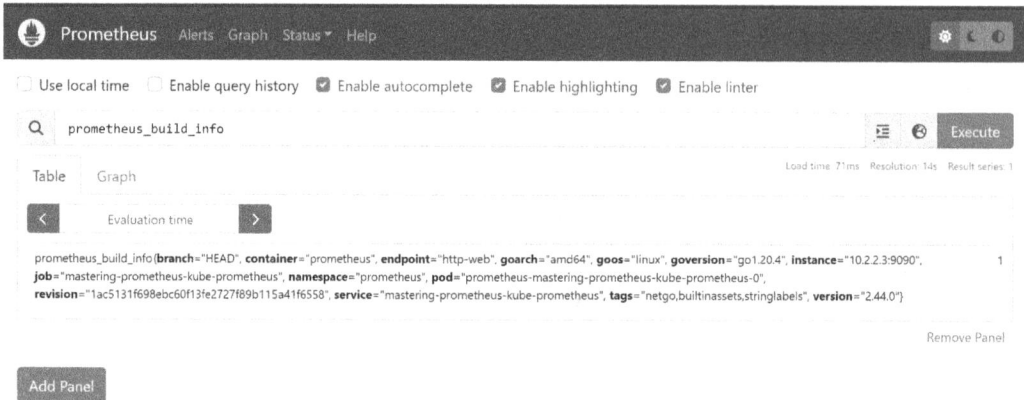

Figure 3.8 – Prometheus instant query

Range vectors, on the other hand, retrieve data over a given time *range*. This time range is specified as a duration followed by a unit, such as 5m for the past 5 minutes.

The following are valid units for a duration:

- ms: Milliseconds
- s: Seconds
- m: Minutes
- h: Hours
- d: Days (assumes a day is always 24 hours)
- w: Weeks (assumes a week is always 7 days)
- y: Years (assumes a year is always 365 days)

These range vector selectors are often more useful when aggregating data to perform an analysis to derive results such as the number of requests per second a service experienced over a particular time range.

The easiest way to reason about the distinction between an instant vector selector and a range vector selector is to recognize that if the selector includes brackets ([ ]), then it's a range selector.

Instant versus range queries

To make matters slightly more confusing, Prometheus also distinguishes between instant and range *queries*.

Prometheus has two primary API endpoints that are used to retrieve time series data: /api/v1/query and /api/v1/query_range. The former will execute your vector selector by looking at data based on the current time, using the most recent sample available for the time series you select. The latter requires a start time and an end time as URL query parameters because it will execute your selector but over a given time range by re-evaluating the selector at a given step, which is also passed as a query parameter. It is also the primary API endpoint that you'll be using (knowingly or not) since it is what is used for graphing time series data. In my experience, you will rarely *directly* use the API, but rather use a variety of tools that interface with these API endpoints such as Grafana or the Prometheus CLI tool, `promtool`.

> **Important note**
>
> Both instant vectors and range vectors are valid for the `/query` endpoint, but *only* instant vectors (or, technically, scalars) are valid for the `/query_range` endpoint.

For range queries, the step is the interval that determines how often your instant vector is re-evaluated to provide a data point. For example, a query with a step of 30 would re-evaluate the selector every 30 seconds like so until it hits end:

```
$now, $now - (30s * 1), $now - (30s * 2), ...
```

Offsets

Offsets are a nifty feature of Prometheus that allows you to pull time series data from the past based on a time offset defined relative to the current time.

The same duration units from range vectors are usable here. So, to pull a metric's value from 5 minutes ago, you could append `offset 5m` to your instant vector selector.

For example, this query would return the value of `my_cool_custom_metric` from 5 minutes ago:

```
my_cool_custom_metric offset 5m
```

The same syntax is valid for range vector selectors:

```
rate(my_cool_custom_metric[5m] offset 1h)
```

Note that the offset must be directly adjacent to the vector selector. This means that this query would be **invalid**:

```
rate(my_cool_custom_metric[5m]) offset 1h
```

Additionally, offsets support looking *forward* by using a negative offset. In practice, this is rarely used but can still prove useful:

```
rate(my_cool_custom_metric[5m] offset -1h)
```

Subqueries

Subqueries are a hidden gem of PromQL and one of my absolute favorite power user features. Have you ever wished that you could look at an instant vector over a range – for example, "What's the highest the request rate has been over the past hour?" Well, you can!

```
max_over_time(rate(http_requests_total[5m])[1h:1m])
```

Notice the second time range of [1h:1m]? That's all that's needed for you to make a subquery.

Subqueries are specified in [<range>:<resolution>] format. You can think of its resolution like step, which we discussed previously. In the preceding example, Prometheus will evaluate rate(http_requests_total[5m]) every 1m over the past hour. These 60 data points will then be passed into the max_over_time() aggregation over time function, which requires a range vector as an argument.

Additionally, resolution is optional and can be omitted. Here's an example:

```
max_over_time(rate(http_requests_total[5m])[1h:])
```

The colon is still required, but the global evaluation interval in the Prometheus config file (which defaults to 1m) will be used if a resolution is not provided.

Staleness

Metric staleness is one of the most common "gotchas" that I've observed people get hung up on. It's generally the reason behind gaps in graphs, which generates a lot of questions.

When stale, Prometheus metrics can be observed to "disappear." Additionally, they can disappear if no sample is received within the past 5 minutes, which is the default query lookback delta. This means that during querying, Prometheus will look back 5 minutes from a given timestamp trying to find a data point. If none is found, the series looks to disappear during that interval.

> **Configuring a query lookback delta**
>
> While 5 minutes is the default value for how far back a query will look for a sample value, it can be configured using the --query.lookback-delta flag. In practice, this should only be modified with good reason. For example, I have modified it in the past when a Prometheus instance has SNMP scrape jobs with long scrape intervals (15 minutes), during which a system would appear to have no metrics for 10 minutes (15 minutes between scrapes – a 5-minute lookback = 10 minutes of no samples being returned).

Staleness markers are also inserted when a metric that was on the previous scrape is not in the current scrape, which also causes disappearing metrics. This is another reason why it is important to ensure your systems that emit Prometheus metrics consistently emit the same metrics, even if their value is 0.

Series are also marked stale if the target that they were coming from goes away (for example, if a Kubernetes pod is deleted and therefore no longer monitored). This ensures that the last sample value displayed for a series roughly aligns with the time it was removed from Prometheus's list of scrape targets.

Query operators

While it can be great to just pull a single time series, Prometheus shines when you use it to combine multiple different series in your results. This can be done by applying arithmetic across series, comparing them, or joining labels of distinct series together, such as by using `JOIN` in SQL.

Arithmetic

Prometheus supports the following arithmetic operations:

- `+` (addition)
- `-` (subtraction)
- `*` (multiplication)
- `/` (division)
- `%` (modulus)
- `^` (exponents/powers)

In practice, I've observed the division operator to be the most common when comparing two time series. For example, dividing a metric representing the number of requests to an API that returned errors (status code more than or equal to `500`) by the total number of requests to compute the error rate of the service over its lifetime:

```
sum(myservice_http_requests_total{status=~"5[0-9]{2}"})
/
sum(myservice_http_requests_total)
```

Aggregations

Aggregation operators operate across multiple time series to provide data such as the `sum` value of the values of the time series, the `count` value of them, the average/mean (`avg`) across them, or the `max` value of them.

> **Note**
>
> For a full up-to-date list of possible aggregations, please consult the Prometheus project's documentation at `https://prometheus.io/docs/prometheus/latest/querying/operators/#aggregation-operators`.

Aggregation operators are designed to work across multiple time series to be effective. While not strictly necessary, they tend to be used with one or more dimensions to aggregate by (think GROUP BY in SQL). This can be specified using either the `by` keyword or the `without` keyword.

The `by` keyword operates by aggregating all of the matched series based on the provided label(s). For example, to get a count of pods in each namespace, you can use the following command:

```
count(kube_pod_info) by (namespace)
```

This results in the following output:

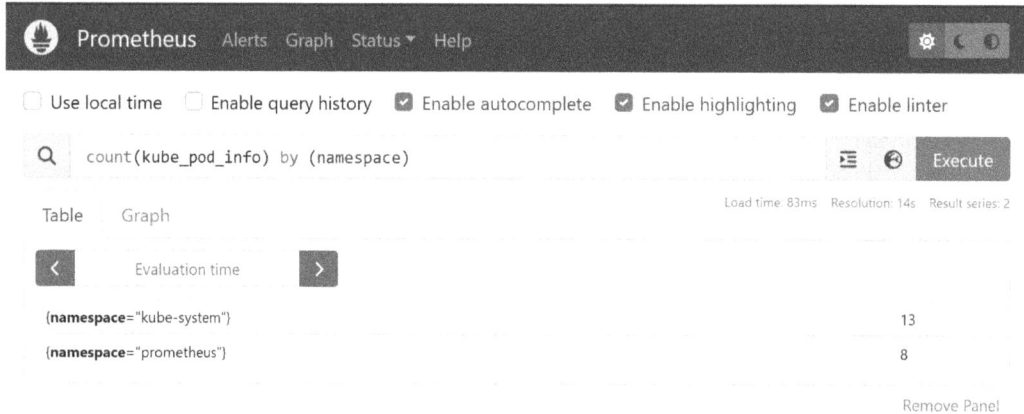

Figure 3.9 – Using the "by" operator

Both `by` and `without` support multiple labels as arguments, so we can also view the count of pods on each node in our cluster from each namespace, like so:

```
count(kube_pod_info) by (namespace,node)
```

This returns the following output:

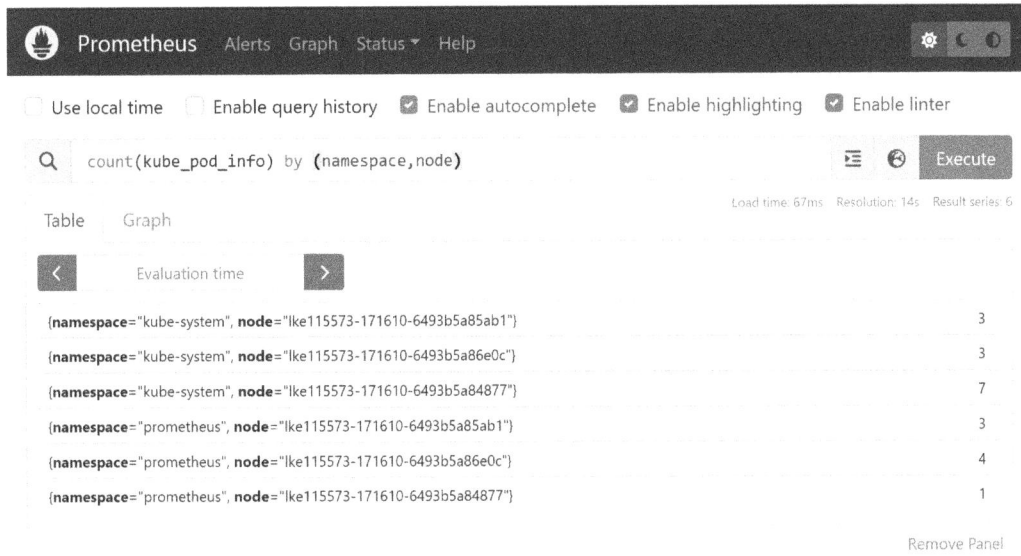

Figure 3.10 – Using the "by" operator with multiple arguments

The caveat with the by operator is that you lose out on a lot of extra data from other labels, which is where the without operator comes in. The without operator will aggregate time series by all unique label combinations in the matched time series *except* the labels that are passed as arguments to the operator.

For example, we could get the number of containers running in our cluster in each namespace using the without operator to remove unique labels other than namespace while preserving all the other present labels, like so:

```
count(kube_pod_container_status_running) without (uid, pod, container)
```

This returns the following output:

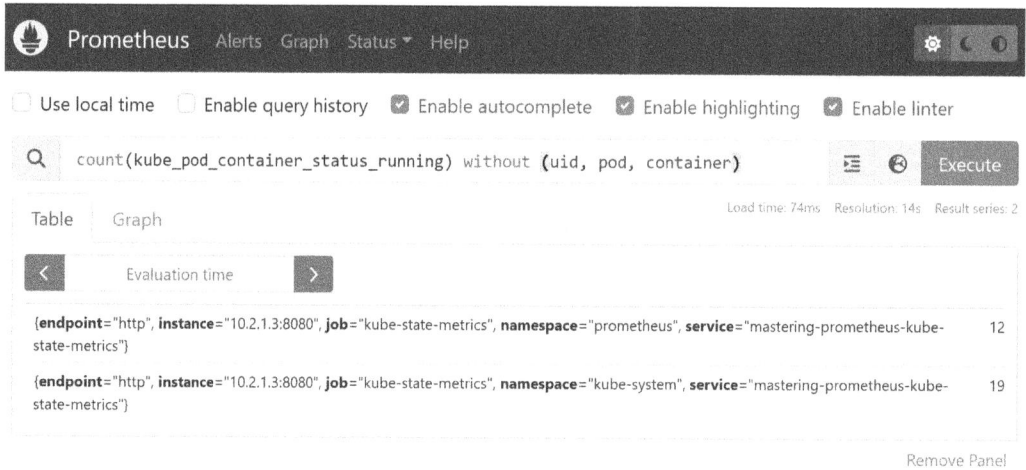

Figure 3.11 – Using the "without" operator with multiple arguments

Generally, you should lean toward using the `without` keyword whenever feasible because it will preserve as many labels as possible, which gives you greater insights into your data.

Group modifiers and vector matching

Prometheus also provides ways to join the labels of different metrics together in the results returned by queries when using the `group_left` and `group_right` query modifiers. These allow for "many-to-one" and "one-to-many" matching of vectors, respectively. If you're familiar with SQL, this is somewhat akin to the `JOIN` command but with time series instead of tables and labels instead of columns.

The direction in the keyword indicates which direction the labels should be applied – either adding labels to the left (take labels from the second series and move them "left" to the first series) or the right (take labels from the first series and move them "right" to the second series).

These modifiers also accept an optional list of label names to move from one side of the query to the other. If not provided, then no labels are added but many-to-one/one-to-many matching is still enabled, which can be desirable.

For this to work, the series that the labels are going to must only have one match from the series the labels are coming from. This is referred to as "many-to-one" matching (`group_left`) or "one-to-many" matching (`group_right`). To coordinate this matching between series, Prometheus provides the `on` and `ignoring` keywords.

The on and `ignoring` keywords work similarly to the `by` and `without` keywords and are **required** when using the grouping operators. Both keywords accept one or more label names as parameters. The on keyword specifies the label(s) that will be used to correlate series from one side of the query to the other. Conversely, `ignoring` specifies which labels to ignore when correlating the series – implying that all the labels that are *not* ignored should match from one side to the other.

Often, you will commonly see the `group_left` and `group_right` operators being used with informational-style metrics that we talked about when we covered gauges. Their whole purpose is to provide additional information about a system. Since they have a static value of 1, we can safely use them in queries as something we can multiply by since it will not change the value of the other metric we are multiplying it with.

Here's an example of how we can combine two of those informational metrics to add information about a Kubernetes node (Kubernetes version and operating system version) to a metric containing information about a pod by matching the pod to the node it's running on:

```
  kube_pod_info{pod=~".*node-exporter.+"}
* on (node) group_left (kubelet_version, os_image)
  kube_node_info
```

Here's the output:

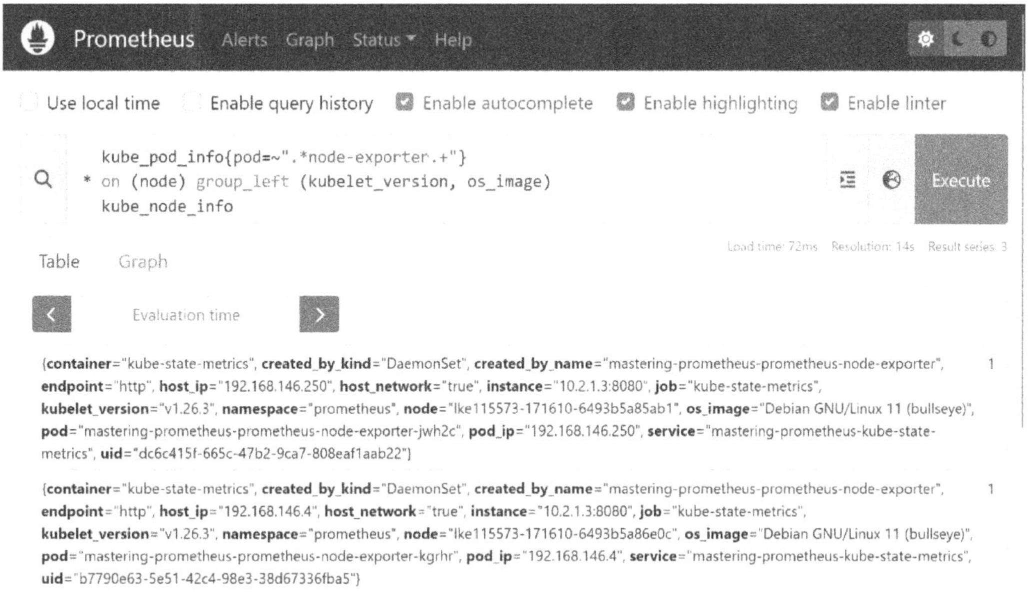

Figure 3.12 – Using group_left

The on and `ignoring` keywords can also be used *without* a grouping modifier to do things such as arithmetic across time series with differing label sets. Without the grouping modifier, the query must return one-to-one matches. For example, to see how much memory Kubernetes pods are using on a node versus how much memory the node has, we can use the following code:

```
sum by (node) (container_memory_rss)
/ on (node)
machine_memory_bytes
```

Logical/set binary operators

PromQL also includes methods for combining and comparing instant vectors using the `and`, `or`, and `unless` query operators.

The `and` operator is the easiest to understand as it is a simple Boolean AND function:

```
vector1 and vector2
```

It requires that for every series in `vector1`, there must be an exactly matching label set in `vector2` (besides the __name__ label). If there is no match for the label set of a series in `vector1` inside of `vector2`, then that series is not included in the results.

Similar to the grouping modifiers, we can use both `on` and `ignoring` to change which labels we are matching between `vector1` and `vector2`. For example, this query would not return anything without specifying the label used to associate the different time series:

```
  node_os_info
and on (pod)
  kube_pod_info
```

The `or` operator works by returning all elements from `vector1` and also all elements from `vector2` that do not have label sets that match series from `vector1`. It's essentially a way to return results from multiple time series in a single query, like so:

```
  node_os_info
or
  kube_pod_info{pod=~".+node-exporter.+"}
```

In the example environment from *Chapter 2*, this query will return six series – three for node_os_ info and three for kube_pod_info.

The `unless` operator is my personal favorite as it has some solid use cases in alerting. It functions like an XOR logical operator, where only `vector1` or `vector2` can return data. If a label set is present in `vector1` *and* `vector2`, then the series will not be included in the results.

For example, kube-prometheus includes an alerting rule (`InfoInhibitor`) that fires when there are alerts with an `info` severity unless there are `warning` or `critical` severity alerts also firing in that namespace. This is done to prevent noise from `info` level alerts, which are generally unhelpful unless there is a larger problem occurring, in which case a higher severity alert would fire and the `info` alerts could assist in pinpointing a cause:

```
  ALERTS{severity="info"} == 1
unless on (namespace)
  ALERTS{alertname!="InfoInhibitor", alertstate="firing",
severity=~"warning|critical"} == 1
```

Query functions

70 different query functions are supported in PromQL at the time of writing, and to go over each of them, even in passing, would take a full chapter on its own. However, there are only a few you need to be aware of for 90% of use cases, so let's go over some of my favorites.

rate/irate

Both `rate` and `irate` are query functions that compute the per-second average increase of a time series within a given *range vector*. However, they do so in slightly different ways.

> **Disclaimer**
>
> Both `rate` and `irate` should only be used on counter and histogram-type metrics. It expects series to only be increasing in value over time.

`rate` is the more traditional way that you would think of computing an average over time: divide the increase from start to the end by amount of time that passed. Here's an example:

```
  rate = (v100 - v0) / (t100 - t0)
```

While `rate` calculates using all data points in the given range vector, `irate` calculates the per-second increase by looking at the two most recent data points. So, even though you may specify some range with 100 samples in it, it would only compute this:

```
  irate = (v100 - v99) / (t100 - t99)
```

Consequently, your `irate` over a 7-day range is almost always going to be the same as an `irate` over a 5-minute range. As such, it's generally recommended to use `rate` in most cases and only leverage `irate` on volatile series that change quickly.

For example, we could use `rate` to see how many samples we're ingesting in Prometheus every second based on how many samples we've added in the past 5 minutes:

```
rate(prometheus_tsdb_head_samples_appended_total[5m])
```

increase

The `increase` function works like the `rate` function but without dividing by elapsed time – it just shows how much the series value went up over a given time range. Consequently, the same caveats apply that it should only be used for counter and histogram-type metrics.

It can be useful for metrics where measuring per-second increases isn't meaningful. For example, we can use it to see how many TSDB compactions Prometheus has run in the past day:

```
increase(prometheus_tsdb_compactions_total[24h])
```

changes

The `changes` function returns a value representing the number of times that a time series' value changed within a given time range. It can be used on all metric types but is most useful on gauges.

I often see it used with series that have a timestamp value (in seconds since the Unix epoch) for the series to track things such as how frequently a service is restarting.

For example, we could get a rough idea of how often pods restarted in the past day with the query:

```
changes(kube_pod_start_time[24h])
```

I say "rough" idea because a caveat is that the pod may have restarted one or more times since the last time Prometheus scraped it. Consequently, when using this function, keep in mind that a value can only change once per scrape interval and therefore may not provide a full picture.

delta

The `delta` function is like the `increase` function but for gauges. Since gauges can go up or down, we shouldn't use the increase function on it. Instead, we can use the `delta` function, which can return negative values if the metric decreases over the given time range.

I've used this in the past to detect silent failures such as when we see a huge dropoff in the number of servers we're monitoring. To accomplish something similar in Prometheus, we could use this query to see the delta of the number of scrape targets Prometheus has in the past day:

```
delta(count(up)[24h:])
```

These are just a few of the query functions available for use in Prometheus. For the extensive list, visit the Prometheus docs page at `https://prometheus.io/docs/prometheus/latest/querying/functions/`.

Summary

In this chapter, we ramped up our learning by looking at Prometheus's data model, reviewing some PromQL basics, and diving deep into the internals of Prometheus's TSDB. I learned a lot and I hope you did, too.

We'll continue to build on our PromQL skills throughout this book, especially in our chapter on alerting. Before we get to that, though, we'll look at one of Prometheus's "killer" features that make it so well suited to containerized and cloud-native environments: service discovery.

Further reading

To learn more about the topics that were covered in this chapter, take a look at the following resources:

- *Ganesh Vernekar's outstanding blog post series on the Prometheus TSDB*: `https://ganeshvernekar.com/blog/prometheus-tsdb-the-head-block/`

- *Prometheus documentation*: `https://prometheus.io/docs/`

- *The original Prometheus TSDB write-up by its author*: `https://web.archive.org/web/20210622211933/https://fabxc.org/tsdb/`

4

Using Service Discovery

In *Chapter 1*, we discussed some of the things that set Prometheus apart from other monitoring tools. One of the major features that distinguishes Prometheus is its built-in functionality around what it calls service discovery. Now we get to learn more about what it is and how it works.

Being able to leverage service discovery and even write custom service discovery mechanisms will allow you to establish Prometheus environments that are truly dynamic and cloud-native. With that in mind, we're going to dive into a comprehensive look at what service discovery is and how you can make the most of it.

In this chapter, we're going to cover the following main topics:

- Service discovery overview
- Using service discovery in a cloud provider
- Custom service discovery endpoints with HTTP SD

Let's get started!

Technical requirements

For this chapter, you'll need the following:

- The Prometheus environment from *Chapter 2*
- Go (>=1.20)

This chapter's code examples are available at `https://github.com/PacktPublishing/Mastering-Prometheus`.

Service discovery overview

In traditional monitoring environments, administrators either needed to know what they would be monitoring in advance or needed to configure push-based monitoring systems. Since the servers and devices being monitored did not change frequently, it was acceptable that adding or removing monitoring required configuration changes. However, in modern cloud-native environments, where the number of instances of an application can scale up and down automatically through systems such as Kubernetes' `HorizontalPodAutoscaler`, this can introduce undesirable gaps in monitoring as targets that should be monitored are constantly added, removed, and replaced. Prometheus solves this through its service discovery system.

The Prometheus service discovery system provides over two dozen different methods of dynamically retrieving and adding scrape targets. These range from cloud providers such as AWS, Azure, and Akamai (Linode) to Docker and Kubernetes to more manual setups using file-based or generic HTTP service discovery. When configured, all service discovery providers are regularly updated in the background of the Prometheus process.

Recall *Figure 2.1* from *Chapter 2*, in which scrape discovery was listed as a Prometheus component. The scrape discovery manager gets kicked off during Prometheus' startup and maintains a set of configured discovery providers that continuously refresh at defined intervals. These discovery providers return lists of potential targets and additional metadata on them (in the form of hidden labels) that can be leveraged during relabeling to filter targets and/or add additional labels to them.

Using service discovery

Service discovery does not just happen automatically. Most Prometheus examples use `static_config` to define scrape targets in `scrape_config` in the Prometheus configuration file. Instead, to use service discovery, we need to update the Prometheus configuration file so that we define one or more scrape jobs that leverage a service discovery provider.

Each service discovery provider can be configured within a scrape job within a section named in the pattern of `<type>_sd_configs` – for example, `kubernetes_sd_configs` or `azure_sd_configs`.

Kubernetes service discovery

When using the Prometheus Operator (as we did in *Chapter 2*), these additional scrape configs go into the aptly named `additionalScrapeConfigs` field of the `Prometheus` CRD for all types of service discovery that do not come from Kubernetes. The Prometheus Operator then automatically handles setting up `kubernetes_sd_configs` fields and associated scrape jobs in the configuration file for the `ServiceMonitor`, `PodMonitor`, and `Probe` resources, which are additional CRDs installed by the Prometheus Operator. Let's look at the default ones created by kube-prometheus to get an idea of how they're used.

`ServiceMonitor`, which is created for scraping Node Exporter and installed on all Kubernetes servers by kube-prometheus, looks like this:

```
apiVersion: monitoring.coreos.com/v1
kind: ServiceMonitor
metadata:
```

```
  name: mastering-prometheus-prometheus-node-exporter
  namespace: prometheus
spec:
  attachMetadata:
    node: false
  endpoints:
  - port: http-metrics
    scheme: http
  jobLabel: jobLabel
  selector:
    matchLabels:
      app.kubernetes.io/instance: mastering-prometheus
      app.kubernetes.io/name: prometheus-node-exporter
```

This tells the Prometheus Operator to create a `kubernetes_sd_configs` section in the Prometheus configuration that will discover and monitor services with labels of `app.kubernetes.io/instance: mastering-prometheus` and `app.kubernetes.io/name: prometheus-node-exporter`. Additionally, it will name the Prometheus job based on the value of a label called `jobLabel` on the discovered service(s) and scrape it via a port defined in the `http-metrics` service.

For reference, the service it discovers looks like this (some of the labels have been removed for brevity):

```
apiVersion: v1
kind: Service
metadata:
  labels:
    app.kubernetes.io/instance: mastering-prometheus
    app.kubernetes.io/name: prometheus-node-exporter
    jobLabel: node-exporter
  name: mastering-prometheus-prometheus-node-exporter
  namespace: prometheus
spec:
  clusterIP: 10.128.244.17
  internalTrafficPolicy: Cluster
  ports:
  - name: http-metrics
    port: 9100
    protocol: TCP
    targetPort: 9100
  selector:
    app.kubernetes.io/instance: mastering-prometheus
    app.kubernetes.io/name: prometheus-node-exporter
  type: ClusterIP
```

All of this culminates in the Prometheus Operator adding a section to the Prometheus config that looks like this (again, this has been significantly truncated for brevity):

```
- job_name: serviceMonitor/prometheus/mastering-prometheus-prometheus-
node-exporter/0
  metrics_path: /metrics
  relabel_configs:
  - source_labels: [__meta_kubernetes_service_label_app_kubernetes_
io_instance, __meta_kubernetes_service_labelpresent_app_kubernetes_io_
instance]
    separator: ;
    regex: (mastering-prometheus);true
    replacement: $1
    action: keep
  - source_labels: [__meta_kubernetes_service_label_app_kubernetes_io_
name, __meta_kubernetes_service_labelpresent_app_kubernetes_io_name]
    separator: ;
    regex: (prometheus-node-exporter);true
    replacement: $1
    action: keep
  - source_labels: [__meta_kubernetes_endpoint_port_name]
    separator: ;
    regex: http-metrics
    replacement: $1
    action: keep
  kubernetes_sd_configs:
  - role: endpoints
    namespaces:
      own_namespace: false
      names:
      - prometheus
```

You can find the full `relabel_configs` section by going to the `/config` endpoint on your Prometheus instance from *Chapter 2* and searching for the `prometheus-node-exporter` job.

A trend you'll notice with the `kubernetes_sd_configs` sections that the Prometheus Operator generates (and for most other forms of service discovery) is how crucial the `relabel_configs` section is – this is where the magic happens. In the preceding example, the actual `kubernetes_sd_configs` section is pretty small – I didn't need to remove anything from it. On the other hand, `relabel_configs` was so long that it could take up several pages on its own. This is because all of the work that is done with metadata coming from service discovery providers needs to take place during relabeling.

Relabeling

In Prometheus, there are two types of relabeling: *standard relabeling* and *metric relabeling*. They both serve distinct purposes and occur at different times in the process of performing a scrape.

Metric relabeling happens *after* a scrape occurs, but before the time series are stored in their chunks. This is when time series can be dropped or modified based on various criteria. We'll cover this in more detail later when we discuss how to address issues of data cardinality within Prometheus.

Standard relabeling occurs *before* a target is scraped and is where we perform additional configuration of our service discovery providers. It performs relabeling on discovered scrape targets to add additional labels and filter the targets we want to scrape.

> **Relabel configuration reference**
>
> Both types are similar in how they are defined. Generally, they take a list of some source labels and perform some `action` on them. The most up-to-date information on what configuration settings are available can be found on Prometheus' documentation site at `https://prometheus.io/docs/prometheus/latest/configuration/configuration/`

In typical Prometheus fashion, what we get from service discovery is a set of labels. All service discovery providers attach a plethora of metadata in various labels prefixed with `__meta_$provider_` that are only available during relabeling. For example, here are the metadata labels that are available during relabeling for Kubernetes endpoints (metadata labels vary by types, such as endpoints, pods, services, and others):

Available meta labels:

- `__meta_kubernetes_namespace` : The namespace of the endpoints object.
- `__meta_kubernetes_endpoints_name` : The names of the endpoints object.
- `__meta_kubernetes_endpoints_label_<labelname>` : Each label from the endpoints object.
- `__meta_kubernetes_endpoints_labelpresent_<labelname>` : `true` for each label from the endpoints object.
- For all targets discovered directly from the endpoints list (those not additionally inferred from underlying pods), the following labels are attached:
  - `__meta_kubernetes_endpoint_hostname` : Hostname of the endpoint.
  - `__meta_kubernetes_endpoint_node_name` : Name of the node hosting the endpoint.
  - `__meta_kubernetes_endpoint_ready` : Set to `true` or `false` for the endpoint's ready state.
  - `__meta_kubernetes_endpoint_port_name` : Name of the endpoint port.
  - `__meta_kubernetes_endpoint_port_protocol` : Protocol of the endpoint port.
  - `__meta_kubernetes_endpoint_address_target_kind` : Kind of the endpoint address target.
  - `__meta_kubernetes_endpoint_address_target_name` : Name of the endpoint address target.

Figure 4.1 – Kubernetes service discovery metadata labels for endpoints (source: https://prometheus.io/docs/prometheus/latest/configuration/configuration/#endpoints)

Recall that `ServiceMonitor` filters for a port named `http-metrics` in discovered endpoints. We can see this take place in relabeling by using the `__meta_kubernetes_endpoint_port_name` metadata label, which is only available during relabeling:

```
- source_labels: [__meta_kubernetes_endpoint_port_name]
  regex: http-metrics
  action: keep
```

By using `relabel_config` like this, we can explicitly keep endpoints with a port that matches `http-metrics`. Implicitly, this means that any discovered endpoints with a port name that *doesn't* match will be dropped. This is how you can filter out irrelevant targets in individual scrape jobs.

In addition to filtering, relabeling helps add labels to discovered targets. Continuing with our example, the name of the scrape job generated by the Prometheus Operator in the configuration file is `serviceMonitor/prometheus/mastering-prometheus-prometheus-node-exporter/0`. That's extremely long and difficult to remember, not to mention unclear. However, when we search Prometheus for metrics with that job label, we get nothing. Why?

```
$ promtool query instant \
    -o json \
    http://localhost:9090 \
    'count({job="serviceMonitor/prometheus/mastering-prometheus-
prometheus-node-exporter/0"})'

[]
```

The reason is that the job label is overwritten during relabeling. We can see this occurring here:

```
- source_labels: [__meta_kubernetes_service_label_jobLabel]
  regex: (.+)
  target_label: job
  replacement: ${1}
  action: replace
```

During this stage of relabeling, Prometheus takes the value of the `jobLabel` label on the discovered service and places it in the `job` label on the Prometheus scrape target. We know from the service definition earlier that that value is `node-exporter`, which *does* return our expected time series:

```
$ promtool query instant \
    -o json \
    http://localhost:9090 \
    'count({job="node-exporter"})'

[{"metric":{},"value":[1690254721.896,"2234"]}]
```

As we can see, there are 2,234 time series with the `node-exporter` value for their `job` label.

Debugging relabeling

Relabeling can be difficult to wrap your head around at first. After all, it's not always clear how a target's label is getting its value set. Thankfully, Prometheus makes it easy to debug relabeling.

To see what metadata labels are assigned to a target *before* relabeling, we can go to the `/targets` page in Prometheus (accessible under the **Status** dropdown in the header). From there, we can hover over any of the targets' labels to see what labels they were assigned before relabeling occurred:

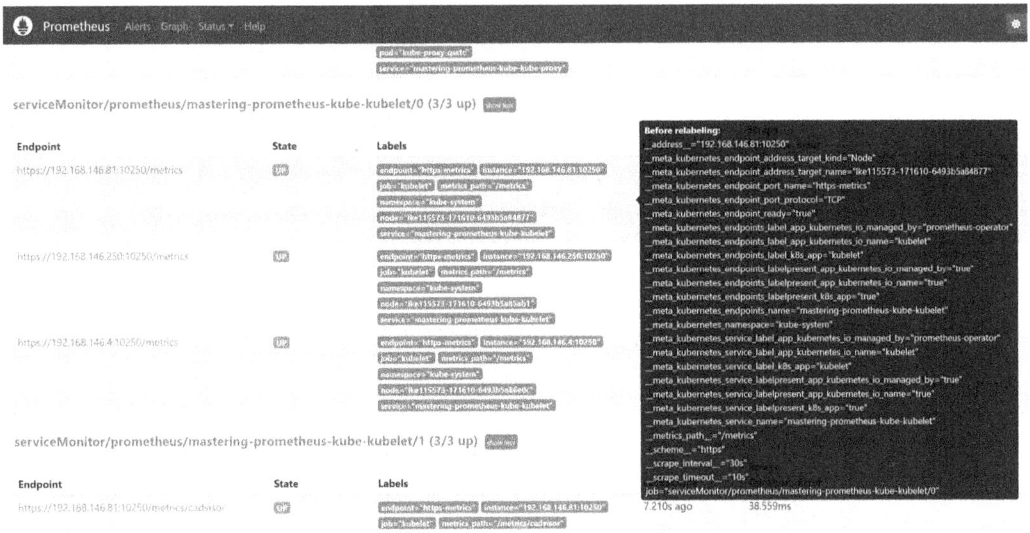

Figure 4.2 – Prometheus' Targets page showing metadata labels
available before relabeling for the kubelet scrape job

Additionally, we can debug Prometheus's service discovery on the `/service-discovery` endpoint, which is also accessible via the **Status** dropdown. This page will show us which targets were discovered in a scrape job and whether they were dropped or not. On top of that, it also shows you the labels before and after relabeling, so it could replace your usage of the `/targets` page for that. In my experience, this page tends to be far more useful for debugging service discovery than the `/targets` page:

serviceMonitor/prometheus/mastering-prometheus-grafana/0 show less

Discovered Labels   **Target Labels**

address__="10.2.2.2:3000"
__meta_kubernetes_endpoint_address_target_kind="Pod"
__meta_kubernetes_endpoint_address_target_name="mastering-prometheus-grafana-6d48958779-ctdj4"
__meta_kubernetes_endpoint_node_name="lke115573-171610-6493b5a85ab1"
__meta_kubernetes_endpoint_port_name="http-web"
__meta_kubernetes_endpoint_port_protocol="TCP"
__meta_kubernetes_endpoint_ready="true"
__meta_kubernetes_endpoints_label_app_kubernetes_io_instance="mastering-prometheus"
__meta_kubernetes_endpoints_label_app_kubernetes_io_managed_by="Helm"
__meta_kubernetes_endpoints_label_app_kubernetes_io_name="grafana"
__meta_kubernetes_endpoints_label_app_kubernetes_io_version="9.5.3"
__meta_kubernetes_endpoints_label_helm_sh_chart="grafana-6.57.3"
__meta_kubernetes_endpoints_labelpresent_app_kubernetes_io_instance="true"
__meta_kubernetes_endpoints_labelpresent_app_kubernetes_io_managed_by="true"
__meta_kubernetes_endpoints_labelpresent_app_kubernetes_io_name="true"
__meta_kubernetes_endpoints_labelpresent_app_kubernetes_io_version="true"
__meta_kubernetes_endpoints_labelpresent_helm_sh_chart="true"
__meta_kubernetes_endpoints_name="mastering-prometheus-grafana"
__meta_kubernetes_namespace="prometheus"
__meta_kubernetes_pod_annotation_checksum_config="291073a4eb8d621f00c54aa27ceb9acd634737fd7ad304f19e209e0df8e695b1"
__meta_kubernetes_pod_annotation_checksum_dashboards_json_config="01ba4719c80b6fe911b091a7c05124b64eeece964e09c058ef8f9805daca546b"
__meta_kubernetes_pod_annotation_checksum_sc_dashboard_provider_config="eeeda0eb284dd445a31b78542fcd35a24a24130c92baa4509b173bf7dba2902f"
__meta_kubernetes_pod_annotation_checksum_secret="1dda38e6fc30fddc689285a1b695556776e8944138db787a8fd1ea43d6689a46"

container="grafana"
endpoint="http-web"
instance="10.2.2.2:3000"
job="mastering-prometheus-grafana"
namespace="prometheus"
pod="mastering-prometheus-grafana-6d48958779-ctdj4"
service="mastering-prometheus-grafana"

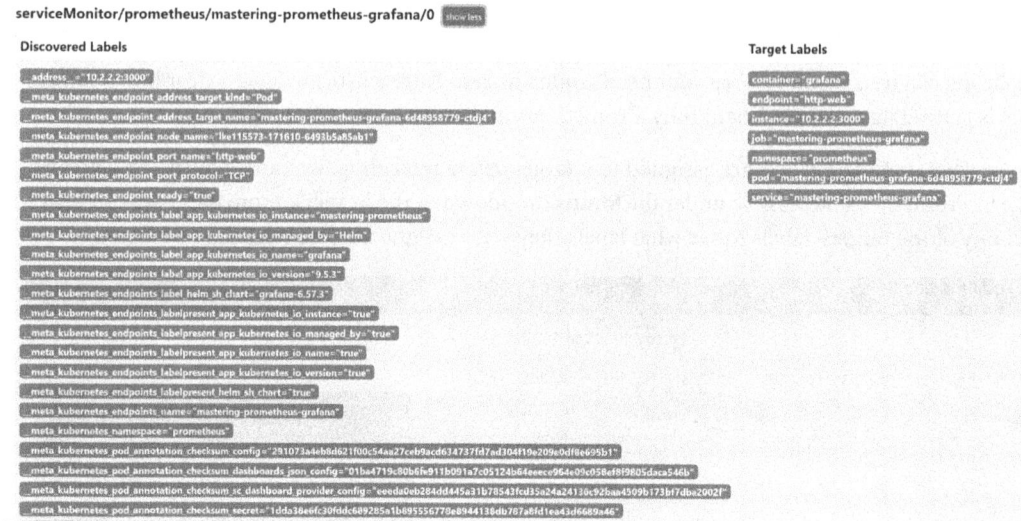

Figure 4.3 – Prometheus' service discovery status page

At this point, I think we can agree that relabeling is where all of the magic truly happens with service discovery to make it useful to an end user. It enables us to broadly retrieve potential scrape targets from an external service such as Kubernetes or a cloud provider and allows Prometheus to filter the targets we want to include. But how do these discovery providers work? Let's take a look at how one cloud provider implements service discovery.

Using service discovery in a cloud provider

While the Kubernetes service discovery we've looked at so far is great and all, you more than likely have some regular old servers that need to be monitored too! How can we discover those? If you're one of the thousands of people running virtualized infrastructure in the cloud, you may be entitled to dynamic service discovery.

Most established cloud providers have service discovery mechanisms built directly into Prometheus for discovering virtualized infrastructure. Such providers include AWS, Azure, GCP, DigitalOcean, Vultr, Hetzner, and – my personal favorite – Linode (also known as Akamai Connected Cloud).

I'm slightly biased in favor of Linode since – at the time of writing – I work there. It also doesn't hurt that a former teammate of mine (TJ Hoplock) wrote and contributed to the Linode service discovery provider, which means I'm pretty familiar with it. So, let's use it as an example of how cloud providers implement service discovery providers for Prometheus.

Linode service discovery

All service discovery providers for Prometheus are defined under the `discovery/` folder in the Prometheus repository. From there, each provider gets its own folder. For example, Linode's service discovery is defined under `discovery/linode/`. This makes sense, right?

> **Important note**
>
> All the code referenced for this section is from the 2.46.0 release of Prometheus.

Most discovery providers are pretty simple and only have one or two `.go` files in their directory. They all implement the `Discoverer` interface, which only has one function – Run:

```
// Discoverers should initially send a full set of all discoverable
TargetGroups.
type Discoverer interface {
        // Run hands a channel to the discovery provider (Consul, DNS,
etc.) through which
        // it can send updated target groups. It must return when the
context is canceled.
        // It should not close the update channel on returning.
        Run(ctx context.Context, up chan<- []*targetgroup.Group)
}
```

However, the way that this interface is implemented may not be what you expect. If you were to look around at different discovery providers, many of them – including Linode – don't define a Run function directly. So, where does it come from for those providers?

In the `refresh` package under `discovery/`, we can see that *it* defines the elusive Run function as a method on the `Discovery` type:

```
// Run implements the Discoverer interface.
func (d *Discovery) Run(ctx context.Context, ch chan<- []*targetgroup.
Group) {
    . . .
}
```

Linode service discovery sets a variable within its own `Discovery` type that references this `refresh.Discovery` type here:

```
        d.Discovery = refresh.NewDiscovery(
                logger,
                "linode",
                time.Duration(conf.RefreshInterval),
                d.refresh,
        )
```

The fourth argument in the call to `refresh.NewDiscovery` is what defines what `refresh` function to call when the Prometheus discovery manager calls the `Run` function. So, we are telling it to run the `refresh` method on our `linode.Discovery` type (d is a pointer to the `linode.Discovery` type).

That `refresh` method we're telling the discovery manager to call is defined within the `linode` package but omitted here due to its length. Essentially, what it does is check if your Linode account has had any **events** since the time the last refresh ran, which is 60 seconds by default. In Linode, an event is any account activity, such as creating a new **virtual machine** (**VM**), deleting a VM, adding an IP address, and so on. If one or more events have occurred, we must trigger a full refresh via the `refreshData` method. This will pull our list of VMs and their associated data from Linode's API and update our targets, which are available via service discovery.

Alternatively, if no events occurred since the last refresh, then we skip pulling an updated list from Linode's API. There is a caveat to this, though: if we go through 10 refresh intervals without any events, we'll still perform that full refresh using `refreshData` just to be safe.

The reason why we poll for events is to limit the number of times we're listing all of our Linodes and IP addresses since that can be an expensive operation (computationally speaking) on the API. In contrast, the `events` API is fairly cheap and quick to return. Doing it this way ensures we only hit those other API endpoints when we need to.

Of course, this behavior is still modifiable. If you create and use a Linode API token that does not have permission to read from the events API, then the Linode discovery provider will intelligently disable event polling and perform a full refresh every time. This is not recommended, but it is an option if you so desire.

When a full refresh occurs, we return a list of pointers to `targetgroup.Group` objects (`[]*targetgroup.Group`), as Run requires. These are the targets that will be available for scraping (presuming they are not dropped during relabeling) with their metadata labels attached. The definition of the `targetgroup.Group` type looks like this:

```
// Group is a set of targets with a common label set(production ,
test, staging etc.).
type Group struct {
    // Targets is a list of targets identified by a label set. Each
target is
    // uniquely identifiable in the group by its address label.
    Targets []model.LabelSet
    // Labels is a set of labels that is common across all targets in
the group.
    Labels model.LabelSet

    // Source is an identifier that describes a group of targets.
    Source string
}
```

Surprise, surprise – just like seemingly everything else in Prometheus, a set of targets is just a set of labels.

The only non-metadata label that's populated in the Linode service discovery provider is the `__address__` label, which is set to the Linode's public IP address plus a port (this defaults to port 80). Any other labels must either be specified in the job or set during relabeling. This is not unique to Linode's service discovery provider, either. The Prometheus code explicitly states that no other labels should be exposed by providers other than the `__address__` label and labels prefixed with `__meta_`.

While service discovery providers such as Linode's are undoubtedly helpful, there are plenty of use cases for service discovery where it might not make sense to write a whole provider to contribute to the upstream Prometheus project. Next, we'll look at how we can implement custom service discovery using a generic, HTTP-based service discovery method.

Custom service discovery endpoints with HTTP SD

For several years after Prometheus' release, a moratorium existed on implementing custom service providers – that is, pull requests to Prometheus to add new service discovery providers would not be accepted. The logic behind this was that it was too much of a burden on Prometheus maintainers to be taking on responsibility for newly implemented service discovery providers and reviewing new submissions when other, more high-priority work could be focused on instead. However, at the end of 2019, that moratorium was lifted.

As expected, the number of service discovery providers increased significantly from 11 before the end of the moratorium to 25 at the time of writing. However, not *everything* should be a service discovery provider in the upstream Prometheus code base. One of the requirements for accepting a new service discovery provider is that wherever you're discovering from needs to be well-established and in use across multiple organizations. Nevertheless, there are plenty of use cases where it'd be nice to have service discovery that connects to some source of truth that is specific to your company or personal needs.

Historically, some folks worked around this by creating some custom service discovery logic that could run as cron jobs and spit out files for consumption by Prometheus' `file_sd` provider. However, this isn't an ideal way to do it – you'd much rather have Prometheus doing the discovery directly rather than some other process that you might not have as much visibility into. Luckily for us, Prometheus maintainer Julien Pivotto introduced a generic `http_sd` type of service discovery in 2021.

This type of generic service discovery is now the recommended method for implementing custom service discovery logic that is not already built into Prometheus. It's dead simple to implement and is done in the same format as a static config (but in JSON, not YAML). Let's build one together.

When building an `http_sd` endpoint, there are a few requirements we need to be aware of:

- Our endpoint must return an HTTP 200 status code
- Our endpoint's response must include a `Content-Type: application/json` header
- Our endpoint's response must be UTF-8 formatted

So long as we check all three of these boxes, Prometheus will be happy with our endpoint.

Next, we need to know the structure that our response needs to be in. As mentioned previously, it matches the format of a static config in the Prometheus configuration but in JSON instead of YAML. So, we may define a static config in YAML like so:

```
- targets:
    - web1.example.com:443
    - web2.example.com:443
  labels:
    component: website
    environment: production
```

We would need to return this in a format that looks like this:

```
[{
    "targets": [
        "web1.example.com:443",
        "web2.example.com:443"
    ],
    "labels": {
        "component": "website",
        "environment": "production"
    }
}]
```

Easy enough. Let's get to it.

> **Note**
>
> I'll use Go throughout this example since it is the language Prometheus is written in. Still, an HTTP SD endpoint can be implemented in any programming language or service that can respond to HTTP requests.

The first thing that we'll need is somewhere to get the data from. For our example, we'll just set up a SQLite database. For our purposes, this can be done within the code, but in the real world, you'll probably already have a database you want to pull from.

To use SQLite, we'll need to install a Go library for it. We can do this with the following command:

```
$ go get github.com/mattn/go-sqlite3
```

Let's start with the barebones of our application in a file I'll call main.go:

```
package main
```

```
import (
    "database/sql"
    "encoding/json"
    "log"
    "net/http"
    "os"
    _ "github.com/mattn/go-sqlite3"
)

var db *sql.DB
var err error


func main() {
    setupDB()
    runHTTPServer()
}

func setupDB() {}
func runHTTPServer() {}
```

Now, let's start defining `setupDB()` by creating a database and defining our table:

```
func setupDB() {
    os.Remove("./http_sd.db")

    db, err = sql.Open("sqlite3", "./http_sd.db")
    if err != nil {
        log.Fatalln(err)
    }
    tableCreate := `
    create table hosts (id integer not null primary key, name text, ip
text);
    delete from hosts;
    `

    _, err = db.Exec(tableCreate)
    if err != nil {
        log.Fatalf("%q: %s\n", err, tableCreate)
    }
}
```

With our table created, we need to populate it with some data. We'll start a database transaction, prepare a statement, and insert a few servers:

```go
func setupDB() {
    os.Remove("./http_sd.db")
    . . .
    tx, err := db.Begin()
    if err != nil {
        log.Fatalln(err)
    }
    tableInserts, err := tx.Prepare(`insert into hosts (name, ip)
values (?, ?)`)
    if err != nil {
        log.Fatalln(err)
    }
    for name, ip := range map[string]string{"web1.example.com":
"192.168.1.39", "web2.example.com": "192.168.1.40"} {
        _, err := tableInserts.Exec(name, ip)
        if err != nil {
            log.Fatalf("%q: Couldn't insert %s %s\n", err, name,
ip)
        }
    }
    if err = tx.Commit(); err != nil {
    log.Fatalln(err)
    }
    log.Println("Bootstrapped database.")
}
```

And just like that, we've got our database all set up! Now, we can move on to implementing the HTTP server that will return data from this database to Prometheus.

We'll just use the built-in net/http package from Go's standard library for this use case. Before we implement the server, though, we need to define what our response is going to look like. To do so, we can create a custom struct type with JSON tags to ensure its data ends up in the correctly named fields when we marshall it to JSON for the response:

```go
type httpSDData struct {
    Targets []string          `json:"targets"`
    Labels  map[string]string `json:"labels"`
}
```

This matches the example data format we looked at earlier, so we should be good to continue. We'll need to implement a handler function to run when Prometheus hits our HTTP SD endpoint. We can start by querying the data we'll need from the database:

```
func runHTTPServer() {
    handler := func(w http.ResponseWriter, r *http.Request) {
        res := []httpSDData{}
        i := 0
        rows, err := db.Query(`select name, ip from hosts`)
        if err != nil {
            log.Fatalf("Couldn't query hosts table: %q", err)
        }
        defer rows.Close()
    }
}
```

Next, we need to iterate through it, updating the list of httpSDData objects we just created in res.

> **The __address__ label**
>
> Behind the scenes, Prometheus uses the hidden __address__ label to determine what hostname or IP address Prometheus connects to. In most cases, the instance and __address__ labels end up being the same because the value of instance is automatically set to the value of __address__ if it's not already defined. However, in advanced use cases, it can be helpful to have an instance label that differs from the address that Prometheus connects to. In our case, the value of the __address__ label is determined by what is inside the targets key in our JSON object.

We're going to explicitly set the instance label to the server's friendly name, while still using an IP address to connect and thereby avoid monitoring outages caused by DNS:

```
func runHTTPServer() {
    handler := func(w http.ResponseWriter, r *http.Request) {
        ...
        defer rows.Close()
        for rows.Next() {
            var host string
            var ip string
            err = rows.Scan(&host, &ip)
            if err != nil {
                log.Fatalln(err)
            }
            res = append(res, httpSDData{})
            res[i].Targets = append(res[i].Targets, ip)
```

```
                    res[i].Labels = map[string]string{"instance": host}
                    i++
                }
            }
        }
```

Now, we have `res` fully populated with the data we want to return to Prometheus. But we still need to marshall it to JSON and set that `Content-Type` header we mentioned earlier. Let's finish our handler function by doing that and writing the JSON data as the response:

```
func runHTTPServer() {
    handler := func(w http.ResponseWriter, r *http.Request) {
        ...
        encRes, err := json.Marshal(res)
        if err != nil {
            log.Fatalln(err)
        }
        w.Header().Set("Content-Type", "application/json")
        _, err = w.Write(encRes)
        if err != nil {
            log.Fatalln(err)
        }
    }
}
```

Phew – the hard part is done. Now, all we need to do is assign this handler function to a URL path and start the server. Both these things can be done with two single lines of code:

```
func runHTTPServer() {
    ...
    http.HandleFunc("/targets", handler)
    http.ListenAndServe("0.0.0.0:8888", nil)
}
```

Voila! With that, we have a fully functional endpoint that Prometheus can use for HTTP service discovery. All we have to do for Prometheus to pull from our endpoint is define this `http_sd` endpoint in a scrape job, like so:

```
- job_name: http_sd_test
  http_sd_configs:
    - url: "http://http_sd:8888/targets"
```

If you're curious, the folder for this chapter in this book's GitHub repository contains a working example of a self-contained Docker environment that will build and run this code and set up a Prometheus to pull from it.

Summary

In this chapter, we looked at service discovery in Prometheus. It's a key feature that differentiates Prometheus and makes it well-suited to the cloud-native world. First, we looked at how service discovery works by evaluating its usage in the Prometheus environment we set up in *Chapter 2*. Next, we saw how cloud providers can implement service discovery providers in the upstream Prometheus code base by looking at how it's done for Linode. Finally, we got our hands dirty by looking at generic HTTP-based service discovery and building an endpoint for it.

In the next chapter, we'll look at how alerting works in Prometheus and dip back into the world of PromQL to see how we can make our alerts better through advanced queries.

Further reading

To learn more about the topics that were covered in this chapter, take a look at the following resources:

- *Prometheus Service Discovery code base*: `https://github.com/prometheus/prometheus/tree/v2.46.0/discovery`

- *Prometheus HTTP service discovery documentation*: `https://prometheus.io/docs/prometheus/latest/http_sd/`

- *Pull Request to Prometheus implementing the http_sd provider*: `https://github.com/prometheus/prometheus/pull/8839`

5

Effective Alerting with Prometheus

Thus far, we've looked primarily at how to get data into Prometheus through scrape jobs, discovering scrape targets, and manually querying data. But no monitoring system is truly useful if you need to constantly check if everything is okay; we need some system running in the background evaluating the state of our systems and alerting us if they're not working correctly. In this chapter, we'll look at how Prometheus achieves that through a combination of its rule subsystem and the separate Alertmanager component.

We'll cover the following main topics:

- Alertmanager configuration and routing
- Alertmanager templating
- Highly available (HA) alerting
- Making robust alerts
- Unit-testing alerting rules

Let's get started!

Technical requirements

For this chapter, you will need to install the following tools that the Prometheus project bundles with Alertmanager and Prometheus releases, respectively:

- **amtool**: `https://github.com/prometheus/alertmanager/releases/tag/v0.25.0`
- **promtool**: `https://github.com/prometheus/prometheus/releases/tag/v2.46.0`

The code used in this chapter is available at `https://github.com/PacktPublishing/Mastering-Prometheus`.

Alertmanager configuration and routing

It is a common misconception amongst users first starting out with Prometheus that Prometheus itself is responsible for getting alerts to their destination, whether that be Slack, PagerDuty, OpsGenie, or another platform. Alertmanager is the oft-forgotten yet critical component that handles getting the alerts coming from Prometheus to their destination. The divide between the two can be confusing at first, but if you're familiar with Prometheus already, then you know that Prometheus evaluates alerting rules in a constant loop and sends them to one or more Alertmanager instances once they meet a specified `for` duration in the alert. From there, Alertmanager determines where the alert information should be sent. Let's look at how Alertmanager accomplishes that.

Routing

Routing is the core of what the Alertmanager does; it processes an alert by walking a tree structure of **routes** to determine what receiver(s) an alert should be sent to. The only requirements for a valid Alertmanager configuration file are a route and a receiver; everything else is optional.

A route—in its simplest form—is a set of one or more **matchers** and a **receiver**. Below is an example of a simple, valid route:

```
- receiver: "default"
  matchers:
    - "environment = production"
```

We'll discuss receivers more later on, but for now, let's look at matchers.

Matchers

Matchers are a list of strings that correspond roughly to how **PromQL** matchers work. Each entry in the list contains a matcher for a label attached to an alert and its value. The primary difference vs. PromQL is that you do not need to quote a label's value. However, it's still valid if you do. This route would do the same thing as the one above:

```
- receiver: "default"
  matchers:
    - environment = "production"
```

> **Beware single quotes**
>
> While it is valid to wrap a label value in double quotes (`"value"`), it is invalid to wrap it in single quotes (`'value'`). When wrapped in single quotes, Alertmanager will expect the literal single quote characters to be present in the label's value. A matcher of `environment = 'production'` would only match an alert with an `{environment="'production'"}` value. So, keep in mind that double and single quotes are not interchangeable in this context and can lead to unintuitive behavior.

In my opinion, it's more readable if you don't quote the label value. Similar to PromQL label matchers, you can use `=`, `!=`, `=~`, and `!~` to compare equality, inequality, regex matching, and inverse regex matching, respectively. For example, see the following:

```
- receiver: "default"
  matchers:
    - "environment = production"
    - "namespace != default"
    - "team =~ (devops|sre)"
    - "service !~ ^(api.+)"
```

If you've been using Prometheus and Alertmanager for a while, you may be familiar with the `match` and `match_re` sections of a route's configuration. However, the `matchers` section in a route's configuration replaces the previously used (but now deprecated) sections of `match` and `match_re`. Consequently, it is not advisable to continue using those deprecated sections for new routes, and it is recommended to convert any existing routes you may have by using them in Alertmanager.

Subroutes

Alertmanager has a single top-level route that matches all alerts coming in (it does not and cannot have a `matchers` section). All other routes are **subroutes** of this top-level route. You may have noticed in the previous examples that each route was a list item in YAML. That's because each of the examples thus far has been a subroute of that top-level route. In reality, the full route configuration would look more like this:

```
route:
  receiver: "fallthrough"
  routes:
    - receiver: "default"
      matchers:
        - "environment = production"
        - "namespace != default"
        - "team =~ (devops|sre)"
        - "service !~ ^(api.+)"
```

Each subroute can have additional subroutes, which is when Alertmanager routing logic can start to get tricky.

Alertmanager will attempt to match the most specific route possible when determining which receiver to use. If it matches a parent route but none of its subroutes, it will just fall back to using the parent route's receiver. For example, take this configuration:

```
route:
  receiver: "fallthrough"
  routes:
    - receiver: "slack-devops"
      matchers:
        - "team = devops"
      routes:
        - receiver: "pagerduty-devops"
          matchers:
            - "environment = production"
```

With the above configuration, alerts would be routed as follows:

Alert Label Set	Matched Receiver
{environment="production", team="devops"}	pagerduty-devops
{environment="testing", team="devops"}	slack-devops
{environment="production", team="sre"}	fallthrough

An exception to this rule is if you enable the `continue` setting in a route. If that is enabled, then an alert will not stop at the most specific subroute it encounters. Instead, it will continue to match subsequent "sibling" routes at the same level as the matched route. This propagates upward as well. Take this extended configuration as another example, where we'll add `continue: true` to the subroute for the `slack-devops` receiver:

```
route:
  receiver: "fallthrough"
  routes:
    - receiver: "slack-devops"
      continue: true
      matchers:
        - "team = devops"
      routes:
        - receiver: "pagerduty-devops"
          matchers:
            - "environment = production"
    - receiver: "default"
      matchers:
```

```
            - "environment = production"
            - "namespace != default"
            - "team =~ (devops|sre)"
            - "service !~ ^(api.+)"
```

Now, alerts would be routed as follows:

Alert Label Set	Matched Receiver(s)
{environment="production", team="devops"}	pagerduty-devops default
{environment="testing", team="devops"}	slack-devops
{environment="production", team="sre"}	fallthrough

Even though we matched the `pagerduty-devops` receiver—which is a subroute of the slack-devops receiver—we continue iterating through the siblings of the slack-devops receiver route since that's where we set `continue` and match another sibling route if possible.

Overall, using `continue` can be exceptionally useful for cases in which you want alerts to go to multiple places (e.g., send a Slack message and an email).

Alert grouping

One of the other core features of Alertmanager is the ability to **group** alerts. We all know what it's like to suddenly be flooded with 100 alerts that are all essentially for the same thing. Well, Alertmanager can help with that by enabling you to group similar alerts based on labels.

Alert grouping can happen at any level within the routing tree and can be different for individual routes. Generally, you will want to configure some default grouping in the top-level route at least. For any subroutes that don't define their own `group_by` setting, they'll just use the grouping settings of their nearest parent.

In the Alertmanager that was deployed via kube-prometheus in *Chapter 2*, the default grouping is achieved just by using the `namespace` label. However, I've typically seen the top-level grouping enacted by using something else, such as the alert name, the service it applies to, and the environment of the service (e.g., "production"). Something that looks like this:

```
route:
  receiver: "fallthrough"
  group_by: ["alertname", "service", "environment"]
```

With this configuration, any alerts that have the same values for those three labels will be grouped. The logic of how this grouping happens can be confusing for some, though, so let's walk through it.

In addition to the `group_by` key, there are also `group_interval` and `group_wait` settings in each route.

The group_wait determines how long the Alertmanager will wait before actually dispatching an alert to its receiver. This is to allow time for additional alerts to come into Alertmanager and be added to the group before it is sent in order to cut down on the number of alerts that are ultimately sent. By default, it is 30 seconds.

The group_wait timer begins once the first alert in an aggregation group is received and is not reset each time that a new alert is added to the group. After the timer expires, the alert is dispatched to its receiver, and any new alerts will wait to be sent until the duration of the next group_interval elapses.

The group_interval setting controls how frequently you want updates to an alert group to be sent. This includes both alerts that are newly firing and alerts that have been resolved. By default, it is 5 minutes. To give an example of how it looks in practice with the default values plus our group_by from above, reference the following table:

Alert Labels	Time Received	Time Sent
alertname="HighAPIErrorRate", service="api", environment="production", instance="api1"	2023-08-04 10:00:00	2023-08-04 10:00:30
alertname="HighAPIErrorRate", service="api", environment="production", instance="api2"	2023-08-04 10:00:17	2023-08-04 10:00:30
alertname="HighAPIErrorRate", service="api", environment="production", instance="api3"	2023-08-04 10:00:35	2023-08-04 10:05:00

The third alert in the group comes in at slightly more than 30 seconds after the first alert, so it misses the initial group_wait time. Instead, it needs to wait until 5 minutes elapse from the time the first alert was received to be sent.

The group_interval only controls how often **updates** to the group will be sent to renotify the folks looking at your alerts, but what about just re-notifying, in general, if no new alerts are added/removed from the group but the alerts still haven't been addressed? For this, Alertmanager has a setting called repeat_interval.

The repeat_interval defines how often an alert group is re-sent to its receiver and defaults to 4 hours. In my experience, this can be a too-aggressive setting and lead to alert fatigue in teams. Consequently, I generally set it to be at least 24 hours. However, it is still a useful feature to keep enabled to ensure that alerts are not flying under the radar or being lost in other noise.

Time intervals

Finally, the last settings we have to talk about concerning Alertmanager route configuration have to do with configuring time intervals for your alerts. This is an optional feature that was only added relatively recently (2022) to Alertmanager, and it enables you to define the time ranges within which you want to receive alerts from a given route. For example, on my team, we only want to receive "warning"-level alerts during business hours and not wake up the on-call engineer to deal with them in the middle of the night. With Alertmanager's time interval features, we can accomplish this.

The two settings that control this are `mute_time_intervals` and `active_time_intervals`. They are not mutually exclusive and can be used in conjunction with each other, but generally, just one is used in a given route. Additionally, they cannot be set on the top-level route in the Alertmanager config.

They are aptly named and consequently self-explanatory in what they do: one defines what time interval(s) to mute alerts, and the other defines the time interval(s) in which to send alerts. If neither is defined, alerts are sent at all times without restriction.

Each configuration setting accepts a list of time interval names that are defined elsewhere in the configuration. To continue the example of only sending alerts during business hours, this is how that would be configured:

```
route:
  receiver: "fallthrough"
  routes:
    - receiver: "slack-devops"
      continue: true
      matchers:
        - "team = devops"
      active_time_intervals:
        - business-hours
time_intervals:
  - name: business-hours
    time_intervals:
      - weekdays: ["Monday:Friday"]
        times:
          - start_time: 09:00
            end_time: 17:00
        location: "US/Eastern"
```

Notice that in addition to specifying `active_time_intervals` (or `mute_time_intervals`) on the route, you must also define a corresponding time interval under the top-level `time_intervals` key in the configuration.

Each time interval can contain one or more ranges of time within it using a variety of settings, including ranges of weekdays, days_of_month, months, years, and the obvious minutes/hours of the day using times. All of these settings are lists and accept ranges using a colon (:) as the delimiter, as per the preceding weekdays example. All time range fields in the interval are optional, but here's an example of what they all look like when in use:

```
time_intervals:
  - name: example
    time_intervals:
      - weekdays: ["monday:thursday", "friday"]
        days_of_month: ["1:31"]
        months: ["1:6", "july:december"]
        years: ["2022:2048", "2049"]
        times:
          - start_time: 09:00
            end_time: 17:00
        location: "US/Eastern"
```

An additional neat feature around the days_of_month setting is that you can use negative numbers to count from the end of the month. For example, -1 in February would correspond to February 28th (or February 29th in leap years), or the range -3:-1 could be used to specify the last 3 days of a given month.

Finally, the location setting in a time interval corresponds to the name of a time zone in the IANA time zone database that should be used as the base for the time interval. For example, "US/Eastern," "Asia/Kolkata," or "Europe/London." Additionally, it accepts "Local" or "UTC" to use the local system time or UTC time, respectively.

Again, using time intervals is entirely optional, and it should be noted that setting time intervals on a route does not mean that the route is entirely ignored when outside of the time intervals. Routing continues to happen normally regardless of time intervals, meaning that it will end the traversal of the route tree if the continue option is not enabled; it is just that notifications might not be sent depending on whether the current time is inside/outside of the time intervals. In my team, I've seen this lead to some confusion when an engineer gets paged for an issue during business hours, resolves it, but sees no "Resolved" notification come through because the time is then outside of business hours, and notifications (including resolved notifications) are no longer being sent.

Receivers

Even though routing is the most critical part of the Alertmanager configuration, receivers are what Alertmanager is really all about. After all, the whole point of the Alertmanager is to deliver alerts to one or more destinations and that's exactly what receivers do—specify how to send to those destinations.

There are a multitude of different receivers in Alertmanager, and we won't be covering them in-depth since the ones you use will largely depend on your individual needs and/or the company's existing technologies. However, suffice to say that each receiver's configuration will vary based on the individual technology. For example, there will not be much overlap in configuration settings for a Slack receiver vs. a Pagerduty receiver.

All receivers get a name and one or more configurations for notifiers such as Slack, Pagerduty, and email. A receiver can have more than one notifier to which it sends alert notifications. For example, you can send to both Slack and Pagerduty (or any other combination of notifiers) like this:

```
receivers:
  - name: "sre"
    slack_configs:
      - channel: "#alerts"
        api_url: "https://hooks.slack.com/services/ZZZZZZZZZZZZZZZZ"
    pagerduty_configs:
      - routing_key: "XXXXXXXXXXXXXXXX"
```

All receivers do have two configuration settings in common: `send_resolved` and `http_config`. `http_config` just affects the client that is used by Alertmanager to connect to a notification destination, which you should very rarely (if ever) need to modify. The `send_resolved` setting is more interesting as it allows you to control whether or not to send notifications when an alert resolves.

By default, all receivers send notifications both when an alert is firing and when it resolves. However, there may be cases when you might just want to know that something happened rather than needing to know that it was resolved as well. An example I've seen enacted is for sending an alert when the Linux **out-of-memory** (**OOM**) killer is triggered. In this scenario, I just want to know that a process was OOM-killed, and I don't need a resolved notification to say that no more OOM kills occurred. In this case, I could set `send_resolved: false` on the receiver to prevent unhelpful resolved notifications from being sent.

Inhibitions

The last section of the Alertmanager configuration we need to cover is **inhibition rules**. Inhibition rules enable you to use Alertmanager to *prevent* notifications from being sent under certain conditions. This is a powerful tool to help minimize alert fatigue and excessive noise.

Alert Fatigue

We've all been there. An alert goes off again. The last five times the alert fired, it's just been a false alarm, so you start to ignore it. This continues until one day, it's actually indicative of an issue, and it gets ignored. This is a form of alert fatigue from low-quality alerts. However, there's another form of alert fatigue that comes from excessive alerts.

Rarely in our line of work do issues happen in a vacuum. You don't need an extra alert to tell you that `server1` in datacenter Y is having connectivity issues when you've already got an alert saying that datacenter Y is experiencing a network outage because, of course, `server1` is going to be having issues.

Inhibition rules work by **inhibiting** other alerts from firing. Essentially, "Don't send me AlertB when I already know AlertA is firing." These are deceptively simple to set up in Alertmanager but are a bit tricky to get right. They look like this:

```
inhibit_rules:
  - source_matchers:
      - severity="critical"
    target_matchers:
      - severity=~"warning|info"
    equal:
      - namespace
      - alertname
```

In inhibit rules, a source is an existing alert, and the target is an incoming alert. So, in the above example, a target alert coming in with the labels `{namespace="default", alertname="HighCPU", severity="warning"}` would be inhibited by another alert with the labels `{namespace="default", alertname="HighCPU", severity="critical"}`. Thus, the warning severity alert would not be sent to the receiver, only the critical severity one. This is a common way of using inhibition rules, where multiple thresholds for alerts are set up, such as a warning when disk usage is 90% and a critical alert when it's 98% full.

Important note

Alertmanager has built-in safeguards to prevent an alert from inhibiting itself. If an alert matches both sides of an inhibit rule (the target side and the source side), then the inhibition is not applied. Ideally, though, you should choose target and source matchers that prevent this case from ever occurring to save yourself the headache of trying to figure out why your inhibit rules aren't behaving intuitively.

The inhibition rules are applied during the **notify** stage of the Alertmanager pipeline, so as long as both alerts arrive prior to a notification going out, then the inhibition will apply (even if, for example, the critical alert arrives after the warning alert for some reason).

Validating

With all the various configuration options we've gone over for the Alertmanager thus far, it's clear that there's a likelihood that you may write an invalid configuration file at some point. Plus, it's YAML – so we're all bound to make mistakes with too much or too little whitespace. Thankfully, the `amtool` binary that ships with Alertmanager has built-in ways to validate and test our configurations.

You can validate that there are no errors in your configuration by running the following:

```
$ amtool check config alertmanager.yml
```

where `alertmanager.yml` is the path to the file defining your Alertmanager's configuration.

Additionally, `amtool` has the functionality to output the hierarchy of route definitions in your configuration file:

```
$ amtool config routes --config.file alertmanager.yml
```

Uniquely, the commands under the `amtool` config are also able to be run without having the Alertmanager configuration file available locally. Instead, it can be pointed at an Alertmanager URL from which it will be able to determine its configuration and run the desired commands like this:

```
$ amtool config routes --alertmanager.url http://localhost:9093
```

By using this functionality, we can also test what route an alert might go to based on its labels. For example, we can confirm the routes we discussed earlier like this:

```
$ amtool config routes test \
  --config.file alertmanager.yml \
  'environment=production team=sre'

fallthrough
```

Finally, an additional helpful validation step to take advantage of is to use the routing-tree editor tool on the Prometheus website. It provides essentially the same functionality as the `amtool config routes` command but in a prettier web interface available at `https://prometheus.io/webtools/alerting/routing-tree-editor/`.

Now that we fully understand how to configure routes and receivers, it's time we look at how we can control the way that alerts are displayed when they're **sent** to those receivers.

Alertmanager templating

Templating can be one of the most difficult-to-grasp concepts within Alertmanager for those unfamiliar with the **Go** templating language. Prometheus and Alertmanager use Go's templating engine under the hood with some additional convenience functions added in. Consequently, you likely have some exposure to it through writing Prometheus alerting rules already since Go templating is often used to dynamically substitute label and series values in alerts. For example, see the following:

```
annotations:
    description: Configuration failed to load for
                 {{ $labels.namespace }}/{{ $labels.pod}}.
```

Prometheus runs the alert annotation fields through a templating engine that enables that description to be rendered dynamically based on the labels attached to the alert. However, the usage of templating in Alertmanager often extends beyond just substituting a handful of values. Rather, you're templating the entirety of what a Slack message will look like or the description of your Pagerduty incident. When doing this, you're far more likely to be utilizing features such as loops and whitespace management.

A full exploration of the Go templating language is beyond the scope of this book, so let's just focus on some basics.

Configuring templates

Alertmanager discovers templates specified under the top-level `templates` key in the configuration file and accepts a list like this:

```
templates:
    - /path/to/my/template.tmpl
    - /other/path/*.tmpl
```

Files do not need the `.tmpl` file extension, but it is convention to use it.

Without any custom templates specified, Alertmanager will only use its built-in default templates for its various supported receivers. Let's look at one of the most commonly used built-in templates, which is the `__subject` template. This template is used by default in most receivers, including Slack messages, Pagerduty incident titles, and Discord message titles. This is what it looks like:

```
{{ define "__subject" }}[{{ .Status | toUpper }}{{ if eq .Status
"firing" }}:{{ .Alerts.Firing | len }}{{ end }}] {{ .GroupLabels.
SortedPairs.Values | join " " }} {{ if gt (len .CommonLabels) (len
.GroupLabels) }}({{ with .CommonLabels.Remove .GroupLabels.Names }}{{
.Values | join " " }}{{ end }}){{ end }}{{ end }}
```

That's a lot to decipher, and part of the reason why templates are so tricky is because they're not particularly readable when they have so much going on like in this one.

Essentially, how it works is like this:

- The `define` keyword specifies that we are defining a new named template that will be closed with a corresponding `{{ end }}` block.

- Next, some brackets are placed so that they surround a count of firing alerts that would look like [FIRING:4] when rendered, where 4 is the number of alerts firing within that alert group.

- Then, all of the group labels' values are concatenated with a space delimiter, which makes our rendered template look something like this: [FIRING:4] production api us-east-1.

- Finally, we also include any shared common labels across the group. These are labels that all the alerts share but aren't explicitly grouped by in the route's group_by definition.

Defining your own templates

Defining your own templates for Alertmanager is one of the most useful things you can do with your Prometheus setup to improve its usability for the teams utilizing it. Different receivers can use different templates, so I've seen multiple cases where different teams have different preferences for how their alerts are presented to them, and we can accommodate that with Alertmanager.

Say you set up annotations for all your alerts such that they have description annotation attached, and that's all that you want to see in Slack. We can set up an alert template that loops through alerts and displays the description annotation.

We'd define a template named "descriptions" that looks like this:

```
{{ define "descriptions" }}
{{- range .Alerts.Firing }}
[FIRING] - {{ .Annotations.description }}
{{- end }}
{{- range .Alerts.Resolved }}
[RESOLVED] - {{ .Annotations.description }}
{{- end }}
{{ end }}
```

Notice the usage of the – character in the curly-brace blocks: {{- end }}. It is used as a whitespace control character that will strip preceding or trailing whitespace (including newline characters), depending on whether you put it at the beginning, end, or both of a curly-brace block.

We can even use amtool to test what the rendering looks like using some sample data:

```
$ amtool template render \
  --template.glob '*.tmpl' \
  --template.text '{{ template "descriptions" . }}' \
  --template.data alert_data.json
```

The output of the command is then the following:

```
[FIRING] - CPU is elevated on api1 for >30m
[FIRING] - CPU is elevated on api3 for >30m
[RESOLVED] - CPU is elevated on api2 for >30m
```

How you use and configure templates will largely be dependent on your specific use case and which providers you leverage, but hopefully, this has given you some ideas of how you can extend the built-in templates to better suit your needs.

Highly available (HA) alerting

Unlike Prometheus itself, Alertmanager **does** have built-in **high-availability** capabilities and is natively able to run as a cluster. It provides clustering capabilities using a gossip protocol in which information is shared amongst the members of the cluster by specifying peers when an instance of Alertmanager is started up using the repeatable `--cluster.peer` flag. Alternatively, if you don't want to enable clustering, you can set the `--cluster.listen-address` flag to an empty string to disable clustering altogether (not recommended for production deployments). There are a variety of flags for Alertmanager under `--cluster.*`, but these should not need to be modified for most use cases apart from that peer flag.

The gossip between cluster members only applies to certain information, such as silences and whether or not a notification has already been sent to an alert group. Consequently, gossip between Alertmanagers in a cluster is crucial to avoid receiving duplicate notifications for an alert. Once one Alertmanager instance sends a notification for a specific alert group to a receiver, it will tell the others in the cluster that it did so, and they will skip sending a notification of their own to the applicable receiver.

Notably, Alertmanager instances do not gossip information about what alerts they've received. Therefore, it is critical that you configure your Prometheus instances to send alerts to all Alertmanager instances in the cluster and do not put a load-balancer in front of your Alertmanagers that would result in alerts from Prometheus not being sent to all Alertmanagers.

Cluster sizing

The quantity of Alertmanagers you want to run is going to be dependent on your individual needs. Personally, I've never needed to run more than two or three within a cluster. Unlike other clustered technologies that you may be familiar with, such as etcd, Hashicorp Vault, or databases, Alertmanager does not leverage a consensus algorithm, so there's no need to target different levels of strong consistency for quorum. Instead, you can just use simple consistency calculations such as $n-1$ and $n-2$, where the system will continue to operate as long as one or more instances are operational.

Consequently, the quantity of Alertmanagers tends to matter less than their distribution across **availability zones** (**AZs**). For example, if you run two Alertmanager instances for HA, you should target for them to be in two different AZs, three should be in three different AZs, and so on.

Making robust alerts

The ability to make more robust alerts is one of the distinguishing factors of Prometheus vs. traditional, check-based monitoring solutions such as Nagios. It allows you to consider multiple factors when creating alerts. For example, rather than just alerting on high memory usage on a server, you can easily create an alert that will only fire if you have high memory usage and a high rate of major page faults since that is generally a better indicator of a system experiencing memory pressure. The idea is to craft alerts in such a way that you reduce the number of false positives as much as possible so that alerts only fire when real, visible impact is occurring. This is part of a larger discussion on the philosophy of alerting on symptoms vs. causes, which is covered comprehensively in Rob Ewaschuk's excellent document entitled *My Philosophy on Alerting* (linked at the end of this chapter).

Use logical/set binary operators

In order to make robust alerts in Prometheus, you should liberally leverage the logical/set binary operators described in *Chapter 3*: and, or, and unless.

> **Important note**
> Informational metrics are more complex alerts that come in handy, especially with the usage of _info style metrics, which is a pattern in Prometheus in which you create gauge metrics with a suffix of _info and a static value of 1 and attach a bunch of informational labels to the metric. That way, if any of the label values change, you don't have them attached to every time series, which causes a large amount of churn. For example, kube-state-metrics—which was deployed as part of the kube-prometheus stack in *Chapter 2*—exposes a kube_node_info metric with information on the different nodes within a cluster.

As a practical example, you may want to monitor NTP synchronization on a server. However, NTP may be out of sync after a reboot. One way to handle this would be to consider the node's boot time within the alert. A simplified way to accomplish this would be with this PromQL query:

```
node_timex_offset_seconds > 0.05
unless on (instance)
changes(node_boot_time_seconds[30m]) > 0
```

This would prevent the NTP alert from firing for roughly the first 30 minutes after a reboot (plus whatever the for duration is that you configure for the alert).

Use appropriate "for" durations

More often than not, the difference between a noisy alert and a good alert is as simple as increasing the for duration when defining the alert. Don't be afraid to use for durations that are as long as 1 or 2 hours and save those short for durations (0m and 5m) for critical alerts. As a rule of thumb, I tend to create warning alerts with for durations of no shorter than 15m and critical alerts with for durations of no longer than 15m.

Additionally, ensure that your `for` duration is greater than the rule's evaluation interval. For example, a `for` duration of 3m is of no benefit if your alerting rule is configured to only be evaluated every 5 minutes.

Use _over_time functions

When using longer `for` durations, one of the issues you may run into is the `for` state being reset if you experience a scrape failure. Since the query expression in the alert would return no data when the rule is evaluated after a failed scrape, the alert would no longer be in the `pending` state.

To work around this, use the `_over_time` PromQL functions such as `last_over_time`, `avg_over_time`, and `max_over_time`. This will prevent a failed scrape from resetting the `for` state so long as some data is returned within the time range you specify:

Bad

```
probe_success{job="blackbox_ssh"} != 1
```

Good

```
last_over_time(probe_success{job="blackbox_ssh"}[5m]) != 1
```

I don't recommend using a time range of more than a few minutes to ensure that your alert triggering and resolving is as timely as possible while still being robust. In general, `last_over_time` will be the best option for ensuring scrape failures don't reset `for` durations. If using other functions, such as `max_over_time`, keep in mind that it will affect alert behavior. For example, if `max_over_time` is used in the above example, the time series would have to be 0 instead of 1 for 5 minutes before the alert expression evaluates as true, which is additional lag time on top of any configured `for` duration.

Anomaly detection

In discussing advanced alerting methods, I would be remiss not to mention the idea of using Prometheus to perform anomaly detection. While most alerts we define in Prometheus tend to have clear thresholds (e.g., disk usage is >95%, `up == 0`, etc.), there is a clear use case for anomaly detection with Prometheus data, and Prometheus is well-suited to it. Prometheus is far from a full-featured data analysis platform in and of itself, but we can perform simple anomaly detection by using it to calculate things such as standard deviations.

Anomaly detection inherently involves some knowledge of the mathematical field of statistics, which is outside of the scope of this book. However, I would recommend reviewing the excellent blog post and conference talk from the team at GitLab on how to do alerting based on z-scores and/or the seasonality of data, available at `https://about.gitlab.com/blog/2019/07/23/anomaly-detection-using-prometheus/`. They cover the topic far more thoroughly than I could hope to.

Unit-testing alerting rules

Testing alerts is one of the less often discussed features of Prometheus. I've frequently seen alerts simply tested by crafting a query and running it while looking at a time range in which an incident occurred to confirm that the query would detect it. While it's certainly valid to do so, it doesn't really do anything to ensure the alert will keep working as expected when it's inevitably tweaked and tuned. Building out unit tests for your Prometheus alerting rules allows you to specify series, their sample values, and what you expect firing alerts to look like.

Since alerts are defined and evaluated within Prometheus, we use `promtool` to validate and test alerts instead of `amtool`.

Unit testing is distinct from simply validating rules, which can be done with the following command:

```
$ promtool check rules ch5/test-rule.yaml
```

This will simply validate whether or not your rules have appropriate syntax. However, we also want to validate that they'll fire when we expect them to fire. To do so, we use the `promtool test rules` command.

In order to create unit tests for our rule files, we need to create an additional YAML-formatted file that defines what the tests look like. Let's walk through testing the `probe_success` alert we looked at above.

First, we'll create a file that contains that rule called `test-rule.yaml` with the following content:

```yaml
groups:
  - name: chapter5
    rules:
      - alert: SSHDown
        expr: >
          max_over_time(
            probe_success{job="blackbox_ssh"}[5m]
          ) != 1
        for: 5m
        labels:
          severity: warning
          team: sre
        annotations:
          description: "Cannot SSH to {{ $labels.instance }}"
          grafana: https://grafana.example.com/ssh-dashboard?var-instance={{ $labels.instance }}
```

Then, we can create a unit test file that we'll call `unit-test.yaml`.

Within there, we first need to define which rule files we want it to evaluate its tests against. That will be the test-rule.yaml file we just created.

```
# This is a list of rule files to consider for testing.
# Globs are supported.
rule_files:
  - ./test-rule.yaml
```

Next, we define how often the alert should be evaluated:

```
evaluation_interval: 1m
```

Now, we can get into defining our test cases, starting with defining the input time series we want to evaluate against and the interval at which samples are spaced (akin to your scrape interval).

```
tests:
  - interval: 1m
    input_series:
      - series: 'probe_success{instance="server1", job="blackbox_
ssh"}'
        values: "1 1 1 1 0+0x12"
      - series: 'probe_success{instance="server2", job="blackbox_
ssh"}'
        values: "1 1 1 _ 1+0x12"
```

In this configuration, we create 16 minutes' worth of data for server1 and server2. server1 succeeds for the first 4 minutes but fails for the next 12 minutes. server2 succeeds for 3 minutes, misses a scrape (indicated by _), and then continues succeeding. With our alert, we'd expect server1 to start alerting at the 14-minute mark since max_over_time doesn't start returning 0 for it until after the 9th minute, and the for duration of the rule is 5 minutes, which takes us to 14 minutes.

Important note

The values field uses expanding notation to define series values in a more concise way than explicitly specifying each sample value. It can be confusing to grasp at first, but think of it as A+BxC) where a is the initial value, b is how much to increment each sample multiplied by the sample number, and c is the number of samples to add. So, 1+0x12 expands to be 1+(0*1) 1+(0*2) 1+(0*3) … 1+(0*12). You can do subtraction, such as A-BxC, as well as addition, but not multiplication or division.

For full syntax documentation, consult the Prometheus documentation at: https://prometheus.io/docs/prometheus/latest/configuration/unit_testing_rules/.

To confirm our suspicions, we need to define what alerts we expect to be returned and when. We expect no alerts to be firing at minute 9, so we can add a test for that:

```
tests:
  - interval: 1m
    [ . . . ]
    alert_rule_test:
      - alertname: SSHDown
        eval_time: 9m
        exp_alerts: []
```

Now, when running our test, we should see SUCCESS:

```
$ promtool test rules unit-test.yaml

Unit Testing:  unit-test.yaml
   SUCCESS
```

That's great, but we still need to test that it fires when we expect it to at the 14-minute mark, too. Let's add a test for that. We'll need to define all of the expected labels and annotations as well, which gives us the chance to validate that our templating is working as expected:

```
tests:
  - interval: 1m
    [ . . . ]
    alert_rule_test:
      - alertname: SSHDown
        eval_time: 9m
        exp_alerts: []
      - alertname: SSHDown
        eval_time: 14m
        exp_alerts:
          - exp_labels:
              severity: warning
              team: sre
              instance: server1
              job: blackbox_ssh
            exp_annotations:
              description: "Cannot SSH to server1"
              grafana: "https://grafana.example.com/ssh-dashboard?var-instance=server1"
```

Running `promtool test rules unit-test.yaml` again should give us the same success message. Now's a great time to experiment with tweaking values to see how the test responds to it. For example, if we tweak the `eval_time` down to `13m`, we'll see a failure since the alert will not be firing yet.

Writing unit tests can get fairly complex—mainly for the part where you define series—when you're writing robust alerts leveraging data from multiple time series, but it's worth it to know for certain that those complex alerts are going to work the way you expect.

Summary

This concludes our discussions around Alertmanager for now. In this chapter, we looked extensively at how to configure Alertmanager from routes to receivers to inhibitions, how to write custom Alertmanager templates using the Go templating language, how to leverage Alertmanager's built-in HA capabilities, some tips on making your alerts more robust, and, finally, how to write and run unit tests on your alerting rules. With this knowledge, you should be well on your way to a more resilient and effective alerting strategy with Prometheus and Alertmanager. This is also the conclusion of the first part of our journey, in which we focused on the fundamentals of Prometheus.

Next, we'll dive into part two, which focuses on scaling Prometheus, beginning with exploring how we can shard and federate Prometheus and configure it in a highly available way.

Further reading

To learn more about the topics that were covered in this chapter, take a look at the following resource:

- *My Philosophy on Alerting by Rob Ewaschuk*: `https://docs.google.com/document/d/199PqyG3UsyXlwieHaqbGiWVa8eMWi8zzAn0YfcApr8Q/edit#!`

- *Unit Testing for Rules*: `https://prometheus.io/docs/prometheus/latest/configuration/unit_testing_rules/`

- *Prometheus Alerting Documentation*: `https://prometheus.io/docs/alerting/latest/overview/`

Part 2:
Scaling Prometheus

In this part, we'll learn what goes into taking our first steps toward running Prometheus on a larger scale. We'll use Prometheus' built-in features to scale out horizontally to multiple Prometheus instances as well as to turn Prometheus into a highly available system. With added scale comes additional complications, so we'll learn how to debug Prometheus when it's not working as expected. Finally, we'll dedicate time to exploring the exporter ecosystem around Prometheus by focusing on its most popular exporter: the Node Exporter.

This part has the following chapters:

- *Chapter 6, Advancing Prometheus: Sharding, Federation, and HA*
- *Chapter 7, Optimizing and Debugging Prometheus*
- *Chapter 8, Enabling Systems Monitoring with the Node Exporter*

Advancing Prometheus: Sharding, Federation, and High Availability

If you're reading this book, chances are that you already had *some* experience with Prometheus before finding this book. If so, at some point, you've also probably run into the need to scale Prometheus beyond just a single Prometheus instance managing everything. There are a variety of solutions to this problem and in this chapter, we'll cover a few of the built-in ones. As a bonus, we'll look at how to make your Prometheus metrics highly available.

In this chapter, we're going to cover the following main topics:

- Prometheus' limitations
- Sharding Prometheus
- Federating Prometheus
- Achieving high availability (HA) in Prometheus

Let's get started!

Technical requirements

For this chapter, we'll be using the Kubernetes cluster and Prometheus environment we created in *Chapter 2*. Consequently, we'll need the following tools to be installed to interact with them:

- **kubectl**: `https://kubernetes.io/docs/tasks/tools/#kubectl`
- **helm**: `https://helm.sh/docs/intro/install/`

The code that's used in this chapter is available at `https://github.com/PacktPublishing/Mastering-Prometheus`.

Prometheus' limitations

When you first start using it, Prometheus may seem like it can do anything. It's a hammer and everything looks like a nail. After a while, though, reality sets in and the cracks start to show. Queries start to get slower. Memory usage begins to creep up. You're waiting longer and longer for WAL replays to finish when Prometheus starts up. Where did I go wrong? What do I need to do?

Rest assured, you did nothing wrong and you're not alone. Prometheus, like any other technology, has limitations. Some of them are specific to Prometheus and others are limitations of time series databases in general. So, what are they? Well, the two we need to care about the most are **cardinality** and **long-term storage**.

Cardinality

Cardinality refers to the measure of unique values of a dataset. High cardinality indicates that there is a large number of unique values, whereas low cardinality is a small number of unique values.

Examples of high cardinality data are IP addresses, **universally unique identifiers** (**UUIDs**), and timestamps. Low cardinality datasets tend to have a set, pre-determined number of possible values, such as the type of operating systems (Windows, Mac, and Linux) or process status (running, stopped, and so on).

In the context of Prometheus, high cardinality data can be bad and is often the primary cause of performance issues. Due to the way that the TSDB is designed, high cardinality data is much more costly to query and takes up more space in memory and on disk.

This only applies to labels, though – the *values* of time series are expected to have nearly unlimited cardinality (any value that can be represented by a `float64` data type in Go). Label values tend to be the highest cardinality data in Prometheus' storage, especially when their possible values are **unbounded**.

To be unbounded means to have no meaningful constraint on possible values. For example, an IP address is an essentially unbounded data point. Sure, there are technically "only" 4,294,967,296 possible values for an IPv4 address; IPv6 addresses only have a measly 340,282,366,920,938,463,463,374,607,431,768,211,456. The point is that it's a lot more than 100, which is the recommended maximum cardinality for most label values. Even then, most label values should stay under 10 possible choices to keep Prometheus happy – only a select few should have more than that.

> **Good labels**
> A good label looks like something with a small, fixed number of possible values – for example, a `method` label that maps to HTTP methods such as GET, PUT, POST, DELETE, and so on.

Once cardinality starts growing in Prometheus, you'll quickly see your memory usage shoot up as more and more unique time series need to be kept in the head block, and your disk usage will also go up as more unique time series need to be stored on disk. Both of those problems could be solved by throwing more money at the problem and just giving your server(s) running Prometheus more resources. However, you'll still be stuck with slow query times on those high cardinality metrics since the PromQL engine in Prometheus is – at the time of writing – still single-threaded.

We'll discuss cardinality – including how to identify and remediate it – in the next chapter. For now, just be aware of it as a limitation that we need to be on guard against.

Long-term storage

As you may recall from *Chapter 3*, we discussed how each block in Prometheus' TSDB is essentially its own independent database. That's incredibly cool and allows for some neat things, but it can also be a limitation when we're considering the long-term storage of Prometheus data. After all, having to potentially open and examine dozens of "independent" databases for queries doesn't seem that optimal.

Additionally, we need to take into account that – even with compaction – the maximum size of a block in Prometheus is 31 days. Consequently, storing years' worth of data directly on your Prometheus servers doesn't seem like the greatest idea since you'll have at least 12 huge blocks every year.

> **An anecdote on long-term storage**
>
> Even though I don't recommend doing it, I do know of folks who have stored at least 2 years' worth of Prometheus data locally on disk without major issues. So, take my recommendation with a grain of salt and know that just because better options are out there doesn't mean that using vanilla Prometheus for long-term storage is impossible or impractical.

Other solutions in the Prometheus community assist in more efficient and effective storage of Prometheus data in the long term (and we'll discuss some of them later in this book). For now, though, just know that the length of time that you're storing data locally on your Prometheus instances has the potential to consume a lot of resources and not be as performant as you'd like.

Sharding Prometheus

Chances are that if you're looking to improve your Prometheus architecture through sharding, you're hitting one of the limitations we talked about and it's probably cardinality. You have a Prometheus instance that's just got too much data in it, but… you don't want to get *rid* of any data. So, the logical answer is… run another Prometheus instance!

When you split data across Prometheus instances like this, it's referred to as **sharding**. If you're familiar with other database designs, it probably isn't sharding in the traditional sense. As previously established, Prometheus TSDBs do not talk to each other, so it's not as if they're coordinating to shard data across instances. Instead, you predetermine where data will be placed by how you configure the scrape jobs on each instance. So, it's more like sharding scrape jobs than sharding the data. There are two main ways to accomplish this: *sharding by service* and *sharding via relabeling*.

Sharding by service

This is arguably the simpler of the two ways to shard data across your Prometheus instances. Essentially, you just separate your Prometheus instances by use case. This could be a Prometheus instance per team, where you have multiple Prometheus instances and each one covers services owned by a specific team so that each team still has a centralized location to see most of the data they care about. Or, you could arbitrarily shard it by some other criteria, such as one Prometheus instance for virtualized infrastructure, one for bare-metal, and one for containerized infrastructure.

Regardless of the criteria, the idea is that you segment your Prometheus instances based on use case so that there is at least some unification and consistency in which Prometheus gets which scrape targets. This makes it at least a little easier for other engineers and developers to reason when thinking about where the metrics they care about are located.

From there, it's fairly self-explanatory to get set up. It only entails setting up your scrape job in different locations. So, let's take a look at the other, slightly more involved way of sharding your Prometheus instances.

Sharding with relabeling

Sharding via relabeling is a much more dynamic way of handling the sharding of your Prometheus scrape targets. However, it does have some tradeoffs. The biggest one is the added complexity of not necessarily knowing which Prometheus instance your scrape targets will end up on. As opposed to the sharding by service/team/domain example we already discussed, sharding via relabeling does not shard scrape jobs in a way that is predictable to users.

Now, just because sharding is unpredictable to humans does not mean that it is not deterministic. It *is* consistent, but just not in a way that it will be clear to users which Prometheus they need to go to to find the metrics they want to see. There are ways to work around this with tools such as Thanos (which we'll discuss later in this book) or federation (which we'll discuss later in this chapter).

The key to sharding via relabeling is the `hashmod` function, which is available during relabeling in Prometheus. The `hashmod` function works by taking a list of one or more source labels, concatenating them, producing an MD5 hash of it, and then applying a modulus to it. Then, you store the output of that and in your next step of relabeling, you keep or drop targets that have a specific `hashmod` value output.

> **What's relabeling again?**
>
> For a refresher on relabeling in Prometheus, consult *Chapter 4*'s section on it. For this chapter, the type of relabeling we're doing is standard relabeling (as opposed to metric relabeling) – it happens before a scrape occurs.

Let's look at an example of how this works logically before diving into implementing it in our kube-prometheus stack. We'll just use the Python REPL to keep it quick:

```
>>> from hashlib import md5
>>> SEPARATOR = ";"
>>> MOD = 2
>>> targetA = ["app=nginx", "instance=node2"]
>>> targetB = ["app=nginx", "instance=node23"]
>>> hashA = int(md5(SEPARATOR.join(targetA).encode("utf-8")).
hexdigest(), 16)
>>> hashA
286540756315414729800303363796300532374

>>> hashB = int(md5(SEPARATOR.join(targetB).encode("utf-8")).
hexdigest(), 16)
>>> hashB
139861250730998106692854767707986305935

>>> print(f"{targetA} % {MOD} = ", hashA % MOD)
['app=nginx', 'instance=node2'] % 2 =  0

>>> print(f"{targetB} % {MOD} = ", hashB % MOD)
['app=nginx', 'instance=node23'] % 2 =  1
```

As you can see, the hash of the `app` and `instance` labels has a modulus of 2 applied to it. For `node2`, the result is 0. For `node23`, the result is 1. Since the modulus is 2, those are the only possible values. Therefore, if we had two Prometheus instances, we would configure one to only keep targets where the result is 0, and the other would only keep targets where the result is 1 – that's how we would shard our scrape jobs.

The `modulus` value that you choose should generally correspond to the number of Prometheus instances that you wish to shard your scrape jobs across. Let's look at how we can accomplish this type of sharding across two Prometheus instances using kube-prometheus.

Luckily for us, kube-prometheus has built-in support for sharding Prometheus instances using relabeling by way of support via the Prometheus Operator. It's a built-in option on `Prometheus` CRD objects.

Enabling it is as simple as updating our `prometheusSpec` in our Helm values to specify the number of shards.

> **Note**
>
> Specifying shards via the `Prometheus` CRD is still considered experimental at the time of writing.

Additionally, we'll need to clean up the names of our Prometheus instances; otherwise, Kubernetes won't allow the new `Pod` to start due to character constraints. We can tell kube-prometheus to stop including `kube-prometheus` in the names of our resources, which will shorten the names. To do this, we'll set `cleanPrometheusOperatorObjectNames: true`.

The new values being added to our Helm values file from *Chapter 2* look like this:

```
prometheus:
  prometheusSpec:
    shards: 2
cleanPrometheusOperatorObjectNames: true
```

The full values file is available in this GitHub repository, which was linked at the beginning of this chapter.

With that out of the way, we can apply these new values to get an additional Prometheus instance running to shard our scrape jobs across the two. The `helm` command to accomplish this is as follows:

```
$ helm upgrade --namespace prometheus \
    --version 47.0.0 \
    --values ch6/values.yaml \
    mastering-prometheus \
    prometheus-community/kube-prometheus-stack
```

Once that command completes, you should see a new pod named `prometheus-mastering-prometheus-kube-shard-1-0` in the output of `kubectl get pods`.

Now, we can see the relabeling that's taking place behind the scenes so that we can understand how it works and how to implement it in Prometheus instances *not* running via the Prometheus Operator.

Port-forward to either of the two Prometheus instances (I chose the new one) and we can examine the configuration in our browsers at `http://localhost:9090/config`:

```
$ kubectl port-forward \
    pod/prometheus-mastering-prometheus-kube-shard-1-0 \
    9090
```

The relevant section we're looking for is the sequential parts of `relabel_configs`, where `hashmod` is applied and then a `keep` action is applied based on the output of `hashmod` and the shard number of the Prometheus instance.

It should look like this:

```
relabel_configs:
  [ . . . ]
  - source_labels: [__address__]
    separator: ;
    regex: (.*)
    modulus: 2
    target_label: __tmp_hash
    replacement: $1
    action: hashmod
  - source_labels: [__tmp_hash]
    separator: ;
    regex: "1"
    replacement: $1
    action: keep
```

As we can see, for each scrape job, a modulus of 2 is taken from the hash of the __address__ label, and its result is stored in a new label called __tmp_hash. You can store the result in whatever you want to name your label – there's nothing special about __tmp_hash. Additionally, you can choose any one or more source labels you wish – it doesn't have to be __address__. However, it's recommended that you choose labels that will be unique per target – so instance and __address__ tend to be your best options.

After calculating the modulus of the hash, the next step is the crucial one that determines which scrape targets the Prometheus shard will scrape. It takes the value of the __tmp_hash label and matches it against its shard number (shard numbers start at 0), and keeps only targets that match.

The Prometheus Operator does the heavy lifting of automatically applying these two relabeling steps to all configured scrape jobs, but if you're managing your own Prometheus configuration directly, then you will need to add them to every scrape job that you want to shard across Prometheus instances – there is currently no way to do it globally.

It's worth mentioning that sharding in this way does not *guarantee* that your scrape jobs are going to be evenly spread out across your number of shards. We can port-forward to the other Prometheus instance and run a quick PromQL query to easily see that they're not evenly distributed across my two shards. I'll port forward to port 9091 on my local host so that I can open both instances simultaneously:

```
$ kubectl port-forward \
    pod/prometheus-mastering-prometheus-kube-0 \
    9091:9090
```

Then, we can run this simple query to see how many scrape targets are assigned to each Prometheus instance:

```
count(up)
```

In my setup, there are eight scrape targets on shard 0 and 16 on shard 1. You can attempt to micro-optimize scrape target sharding by including more unique labels in the `source_label` values for the `hashmod` operation, but it may not be worth the effort – as you add more unique scrape targets, they'll begin to even out.

One of the practical pain points you may have noticed already with sharding is that it's honestly kind of a pain to have to navigate to multiple Prometheus instances to run queries. One of the ways we can try to make this easier is through federating our Prometheus instances, so let's look at that next.

Federating Prometheus

The elusive "single pane of glass." Everybody wants it. Every software vendor purports to sell it. The dream is to have one place you can go to see all of your monitoring data in just that single location. Sharding Prometheus may seem antithetical to that but through federation, we can still achieve it.

What is federation? Federation is the process of joining together metrics from multiple sources into a central location. It is useful for aggregating your metrics into a centralized Prometheus instance. They may be a 1:1 match of the metrics present in the "lower" Prometheus instances, but they can also have PromQL query functions applied to them to perform series aggregation and consolidation before storage in the "higher" Prometheus instance(s).

> **Should I federate?**
>
> If you're federating because you've sharded your Prometheus instances, you've probably already run into the issue where one Prometheus grew too big and costly to operate. If so, federating into a central Prometheus is just going to repeat the same problems.
>
> Ideally, federation should be used sparingly and with subsets of metrics – and ideally, those metrics are aggregated. A rule of thumb is that if a metric has an "instance" label, you probably shouldn't federate it. Regardless, only federating a subset of metrics is an immediate non-starter for that "single pane of glass" we were talking about since the federated Prometheus instance won't have *all* of your data in it. If you need someone to tell you to skip federating and go straight to setting up something such as Thanos instead, then here's your sign. You'll save time in the long run.

Federation is a built-in feature of Prometheus and it even has a special API endpoint exposed for it at `/federate`. You can scrape it just like any other metrics endpoint, but you'll need to provide at least one URL query parameter of `match[]`. The value of `match[]` needs to be an **instant vector selector** (see *Chapter 3* if you need a refresher). For example, `http://localhost:9090/federate?match[]=up{job="node-exporter"}` would be a valid URL – you'll need to escape the curly braces if you're using cURL, though.

The `match[]` parameter can be specified multiple times, too. There's no need to set up multiple scrape jobs for federating multiple subsets of metrics. Continuing from the preceding example, `http://localhost:9090/federate?match[]=up{job="node-exporter"}&match[]=node_os_info` is also valid. Additionally, recall that metric names are stored in a hidden `__name__` label, which you can also use to filter.

As you can see, this would get gnarly pretty quickly if we were doing it by hand. Thankfully, Prometheus makes it easy to specify these query parameters in a nicer YAML format. A scrape config for federation might look like this in Prometheus's YAML configuration file:

```
scrape_configs:
  - job_name: 'federate'
    scrape_interval: 15s

    honor_labels: true
    metrics_path: '/federate'

    params:
      'match[]':
        - 'up{job="prometheus"}'
        - 'node_os_info'

    static_configs:
      - targets:
        - 'prometheus-shard-1:9090'
        - 'prometheus-shard-2:9090'
```

> **Show honor to your labels**
>
> Note the `honor_labels: true` setting included in that scrape job. It's important to include this on your federation scrape job(s) to ensure that you don't unintentionally overwrite labels on the incoming metrics.

Now, for a practical example, let's set up federation with our two Prometheus shards. Rather than using the kube-prometheus Helm chart, we'll just set Prometheus up manually using the `Prometheus` CRD.

First, we need to define a `ServiceMonitor` instance that will discover our two Prometheus shards and scrape them via the `/federate` endpoint. That looks like this:

```
apiVersion: monitoring.coreos.com/v1
kind: ServiceMonitor
metadata:
  name: federated-prometheus
  namespace: prometheus
  labels:
    app.kubernetes.io/name: "federated-prometheus"
```

```
spec:
  endpoints:
    - path: /federate
      port: http-web
      honorLabels: true
      params:
        'match[]':
          - 'up{job="prometheus"}'
          - 'node_os_info'
  namespaceSelector:
    matchNames:
      - prometheus
  selector:
    matchLabels:
      app: kube-prometheus-stack-prometheus
      release: mastering-prometheus
      self-monitor: "true"
```

Most of this should look familiar based on the example from earlier of how the scrape job looks in vanilla Prometheus. Now, we just need a Prometheus instance to consume this `ServiceMonitor` scrape config.

We'll define a Prometheus that looks for a `ServiceMonitor` value with the appropriate labels and disable the discovery of any other scrape jobs since we only want to use this for federation. For convenience, we'll also reuse the service account we already use for the other Prometheus instances, so we don't need to deal with RBAC on another account:

```
apiVersion: monitoring.coreos.com/v1
kind: Prometheus
metadata:
  name: federated-prometheus
  namespace: prometheus
spec:
  scrapeInterval: 15s
  serviceMonitorSelector:
    matchLabels:
      app.kubernetes.io/name: "federated-prometheus"

  serviceMonitorNamespaceSelector: null
  podMonitorSelector: null
  probeSelector: null
  scrapeConfigSelector: null
  version: v2.44.0
```

```
replicas: 1
retention: 24h
serviceAccountName: mastering-prometheus-kube-prometheus
```

After applying both of those manifests with `kubectl apply -f ./ch6/federation.yaml`, we should now have a pod running named `prometheus-federated-prometheus-0`. We can port forward again to this new Prometheus instance and confirm that we're seeing metrics from the other Prometheus instances:

```
$ kubectl port-forward \
    pod/prometheus-federated-prometheus-0 9092:9090
```

You should be able to query and see results from both Prometheus shards we set up earlier using a query such as `count by (prometheus_replica) ({job=~".+"})`. We can easily distinguish between a metric's federation source via the `prometheus_replica` label. This label is one of the external labels that's set via `external_labels` in the configuration of the other Prometheus instances. It is important to name external labels so that they won't collide with other label names, especially for use cases such as this. This is because Prometheus attaches these external labels to data it sends during certain operations, such as sending alerts and responding to federation requests.

And just like that, we've federated Prometheus metrics from multiple instances into a central location! Before we move on to the next section, though, I have some parting thoughts.

Federation *seems* neat, but it's not a silver bullet. In practice, I've far more frequently seen it used as a stop-gap to a more scalable architecture than it being the scalable solution itself. In other words, before choosing to federate, you should evaluate and think hard about whether this is the proper solution for your use case. 9 times out of 10, it's not.

Although federation does have its use cases, extending Prometheus by using other projects such as Thanos tends to be a better solution for seeing all of your metrics in one place. We'll talk about Thanos in more depth later in this book, but for now, let's talk about how to achieve high availability of your Prometheus monitoring environment, to which Thanos is one solution!

Achieving high availability (HA) in Prometheus

Your monitoring environment needs to be one of your most resilient services. It can be a joke that there's no such thing as 100% uptime, but your monitoring environment should come pretty darn close. After all, it's what you depend on to let you know when your other services aren't achieving their 99.9% uptime goal.

Thus far, we've only used Prometheus in a single-point-of-failure mode. If Prometheus goes down, all of its metrics and alerts go down with it. This gap in visibility and alerting is unacceptable. So, what can we do about it if Prometheus doesn't have built-in HA like Alertmanager? The answer? Duplicate it.

Who watches the watchmen?

With an HA Prometheus setup, you can (and should) configure your Prometheus instances so that they monitor each other. Presuming they're not running on the same physical hardware, unexpected failures should be isolated and you can be alerted to them still. It's even better if you can put them in different availability zones within the same region. This covers 99.99% of issues, but it may still be worth monitoring your monitoring using a SaaS tool to get closer to that 100% coverage.

The commonly accepted way to achieve a highly available Prometheus setup – without using something such as VictoriaMetrics to store the metrics instead of Prometheus – is to run copies of them. It's as simple as deploying an additional Prometheus instance that is scraping the same targets and evaluating the same rules. Sure, it's twice as expensive, but isn't it worth the cost to sleep better at night? Bad example. You may *not* sleep better at night since you're more likely to receive pages because your monitoring system *isn't* down. I digress.

Essentially, you run two (or more) Prometheus instances that are completely identical *except* for one external label that denotes which replica it is. The Prometheus Operator uses an external label named `prometheus_replica`, but you can name it whatever you want, so long as the values of it are different across replicas. This isn't *strictly* a requirement, but it's highly recommended for ease of use.

Importantly, when adding this new external label, we also need to make a small change to how alerts are sent to Alertmanager. Recall that Alertmanager handles deduplication and grouping of alerts based on labels. With an HA Prometheus setup, we'll be evaluating all of the same alerting rules twice and therefore sending the same alerts to Alertmanager from two different places. If we don't make any changes, those alerts won't be deduplicated by Alertmanager because they have unique label sets because of our replica label. The solution to this lies in `alert_relabel_configs`.

We've covered standard relabel configurations and metric relabel configurations, but there are also alert relabel configurations. These relabelings apply before an alert is sent to Alertmanager. We can use them to drop our replica label, which will allow Alertmanager to deduplicate the alerts coming in from all of our replicas:

```
alerting:
  alert_relabel_configs:
    - regex: prometheus_replica
      action: labeldrop
```

If you use the Prometheus Operator, this configuration will automatically be added for you. And since we mentioned the Prometheus Operator, let's take a look at how we can easily achieve this HA design using kube-prometheus. Spoiler: it only takes one new line.

HA via the Prometheus Operator

In our Helm values, we specify the `replicas` setting, which determines how many identical copies of a Prometheus to run. If you're curious, this applies to shards as well. So, if you have two shards and two replicas, you'll have four Prometheus servers running – two for each shard:

```
prometheus:
  prometheusSpec:
    replicas: 2
    shards: 2
```

Reusing our `helm` command from earlier, we can apply it like so:

```
$ helm upgrade --namespace prometheus \
    --version 47.0.0 \
    --values ch6/values-ha.yaml \
    mastering-prometheus \
    prometheus-community/kube-prometheus-stack
```

Now, with `kubectl get pods`, we'll see four Prometheus instances running:

```
NAME
prometheus-mastering-prometheus-kube-0
prometheus-mastering-prometheus-kube-1
prometheus-mastering-prometheus-kube-shard-1-0
prometheus-mastering-prometheus-kube-shard-1-1
```

However, there's a major limitation here: we're missing our single pane of glass again. Now, if a Prometheus in the HA pair goes down, users need to know to go to the other instance instead. This isn't a great user experience, nor does it enable automation since automated processes relying on Prometheus metrics will need to account for retrying a request on all possible Prometheus instances in case one is down.

There are some solutions to this, though. Federation is one that we've already covered. You can federate multiple Prometheus instances and use some PromQL to drop that `prometheus_replica` label from results and only select one result using a function such as `min` or `max`.

Or maybe you put your Prometheus instance behind a load balancer – if any of them fail a health check, they're taken out of the pool. With this, you get a single URL that will load balance between Prometheus instances. You'll need to do some sort of session pinning, though, since queries may return different data in subsequent requests since they could hit different Prometheus instances. Additionally, it doesn't make it easy to see continuous graphs if you do a rolling restart of your Prometheus instances, such as during a version upgrade. There will be gaps in different places, depending on where you get load-balanced.

There's even a variety of options for sending Prometheus metrics from all of your Prometheus instances to remote storage systems such as Grafana Mimir, VictoriaMetrics, M3DB, InfluxDB, and many more. We'll talk about these later in this book, but for now, just know that these solutions generally entail running large-scale, centralized systems.

Finally, there's Thanos – my favorite option. Thanos is another open source project in the Prometheus ecosystem that is also sponsored by the **Cloud Native Computing Foundation** (**CNCF**), just like other popular projects such as Prometheus and Kubernetes. It has many components for a variety of use cases but for our use case of achieving HA while still having a centralized location to query data, we only need the Thanos Sidecar and Thanos Query components.

Thanos Sidecar is a sidecar process that runs alongside your Prometheus servers and its primary function in this use case is to proxy queries coming via gRPC from Thanos Query to Prometheus:

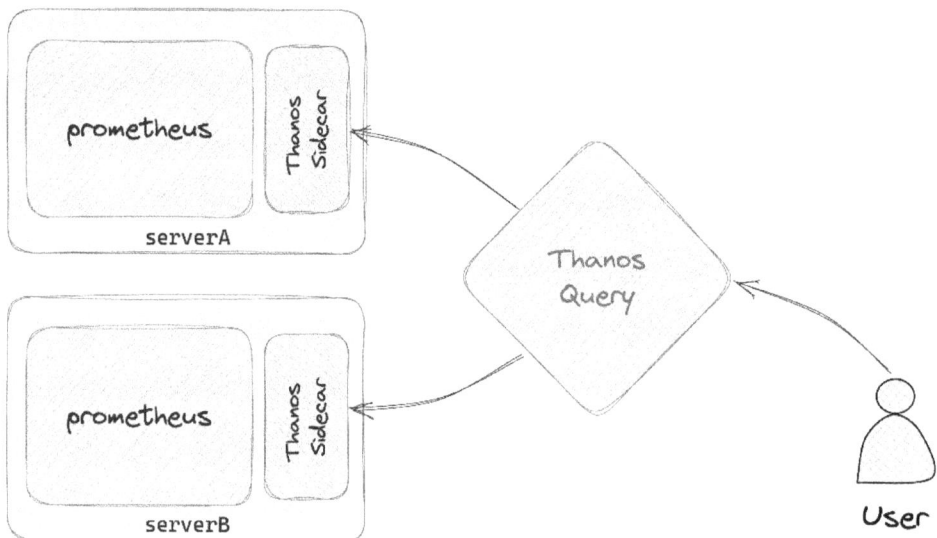

Figure 6.1 – Thanos Query + Sidecar

Thanos Query fans out queries to the relevant Prometheus instances via the sidecar and allows you to deduplicate results from multiple Prometheus instances by defining the name of your replica external label on Prometheus that will be stripped from results. This way, if it sees a gap in data on one replica, it can check the other replica to see if it has data for that time range instead.

We'll talk more about Thanos later in this book, where we have a full chapter dedicated to it. For now, just know that it is a solid, simple option for providing a way to query data from HA Prometheus instances and reduce gaps in graphs.

Cleanup

For future chapters, we won't need the shards of Prometheus that we added in this chapter. You can go ahead and set `shards` to `0` in your `helm values` file and reapply the Helm chart. You may still want to keep the replicas around for later, though, so that you can see how Thanos works with HA Prometheus instances.

Additionally, you can delete the federation Prometheus with `kubectl delete -f ./ch6/federation.yaml` as we will no longer need that.

Summary

In this chapter, we learned how to advance our Prometheus architecture through the use of powerful patterns such as sharding, federation, and highly available replicas. At this point, we have an idea of how to scale our Prometheus instance as our usage grows. But what about when things go wrong? In the next chapter, we'll talk about how we can optimize and debug Prometheus in production.

Further reading

To learn more about the topics that were covered in this chapter, take a look at the following resources:

- *Prometheus TSDB Compaction and Retention*: `https://ganeshvernekar.com/blog/prometheus-tsdb-compaction-and-retention/#condition-2-preset-time-ranges`

- *Prometheus documentation on federation*: `https://prometheus.io/docs/prometheus/latest/federation/#federation`

7

Optimizing and Debugging Prometheus

Even as we scale Prometheus out to multiple replicas and possibly even shards, there will still be performance issues that we need to know how to identify and address. Consequently, this chapter will focus on how we can go about optimizing Prometheus to make the most of the resources it has and how to debug issues when they arise.

In this chapter, we're going to cover the following main topics:

- Controlling cardinality
- Recording rules
- Scrape jitter
- Using pprof
- Query logging
- Tuning garbage collection

Let's get started!

Technical requirements

For this chapter, we'll be using the Kubernetes cluster and Prometheus environment we created in *Chapter 2*. Consequently, we'll need the following tools installed to interact with it:

- **kubectl**: `https://kubernetes.io/docs/tasks/tools/#kubectl`
- **helm**: `https://helm.sh/docs/intro/install/`
- **go**: `https://go.dev/dl/`

This chapter's code is available at `https://github.com/PacktPublishing/Mastering-Prometheus`.

Controlling cardinality

In the previous chapter, we discussed the concept of data cardinality. To refresh your memory, cardinality refers to the measure of unique values in a dataset. We know that time series databases in general have difficulty managing high cardinality datasets without this resulting in severe impacts on query performance.

This cardinality issue isn't even necessarily specific to time series databases – relational databases such as MySQL or PostgreSQL also need to take cardinality into account when selecting data. In relational databases, large tables are often **partitioned** to improve query performance. That's not an option with Prometheus's TSDB, especially since you could consider the data to already be partitioned since each block functions as an independent database. Consequently, the closest we can get to partitioning data based on something other than time is by sharding, as we discussed in the previous chapter.

However, even within these shards, we need to control cardinality to ensure they operate optimally. How do we identify it and remediate it, then? Well, since this is such a common concern for Prometheus users, the project makes it straightforward.

Identifying cardinality issues

The first place to start when identifying cardinality issues is in the web UI of Prometheus. Under the TSDB page (available under the **Status** dropdown, or at `/tsdb-status`), a variety of metrics are available regarding cardinality:

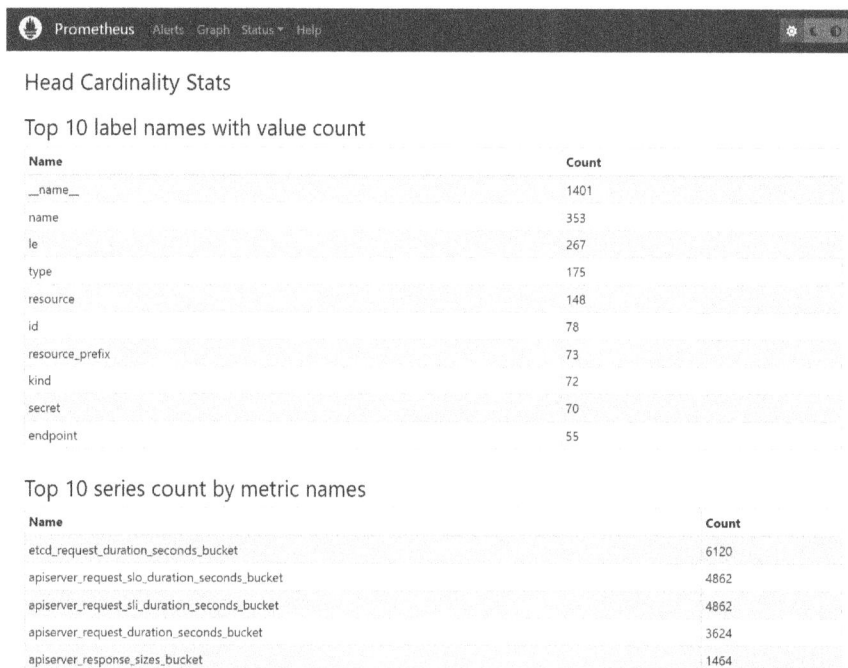

Figure 7.1 – Prometheus's TSDB cardinality statistics

As of version 2.46.0 of Prometheus, this page displays the top 10 statistics for the following:

- Label name with value count
- Series count by metric names
- Label names with high memory usage
- Series count by label value pairs

The `__name__` label is guaranteed to be the label with the most number of values since any other label could only have as many possible values as there are metrics. Consequently, you can ignore it when considering high cardinality labels since it's already presumed to be high cardinality. The label name with value count is going to be the most critical bit of information to look at, but let's explore the others too.

Series count shows which metrics have the greatest cardinality, as measured by the number of individual time series sharing the same metric name. These metrics are going to be slower to query than other metrics in general. High cardinality is often constrained to a few metrics and doesn't necessarily mean that your Prometheus performance will suffer. Instead, in my experience, high cardinality severely negatively impacts querying only specific metrics. The main effect that high cardinality has on the rest of your Prometheus metrics is that the whole instance consumes more resources.

You can see this effect by querying your top metric (`etcd_request_duration_seconds_bucket`, in my case) and observing that it is much slower to return than a lower cardinality metric such as `up` (a 4x difference in my case).

Label names with high memory usage will show you how the cardinality of a single label affects memory usage on your Prometheus instance. It is important to note that this is not only driven by the quantity of values but the *length* of the values as well. Consequently, a label with a cardinality of 200 may consume less memory than a metric with a lower cardinality but longer label values. For this reason, I do not recommend relying on this table to identify *cardinality* issues, but it is still useful to help identify high memory usage caused by labels.

Finally, series count by label value pairs is – in my opinion – the least useful of the four tables. It can be helpful primarily for identifying labels that aren't particularly useful. For example, if every time series has an `environment="production"` label on it, there's not much point in storing it on every time series – either get rid of it or add it as an external label to your whole Prometheus instance. Regardless, this table doesn't relate to cardinality, so let's get back to focusing on how to fix cardinality issues.

Remediating cardinality issues

The tricky part with cardinality is determining where all of these label values are coming from. However, we can use PromQL to identify this without needing to comb through all of our time series manually. Going back to the first table we looked at, let's see what metrics have a name label attached since that's my second-highest cardinality label behind __name__.

Using PromQL, we can see which metrics have the highest cardinality for the name label. This is what the query for this looks like:

```
sort_desc(
  count by (__name__) ({name!=""})
)
```

In my case, the output looks like this:

`kubernetes_feature_enabled{}`	146
`grafana_feature_toggles_info{}`	70
`container_processes{}`	55
`container_memory_max_usage_bytes{}`	55
`node_namespace_pod_container:container_memory_working_set_bytes{}`	55

Table 7.1 – Highest cardinality "name" labels on metrics

As you can see, the `kubernetes_feature_enabled` metric has 146 unique values for the name label. In this case, I don't care about keeping this high cardinality metric around, so I can just drop it.

Using `metric_relabel_configs` in the Prometheus configuration file, we can easily drop specific labels or even entire metrics. Since I don't have a use for this metric currently, I can just drop the whole thing.

> **Be wary of dropping labels**
>
> Using the `labeldrop` action in `metric_relabel_configs` is a powerful way to reduce labels, but it must be used with caution. After applying `metric_relabel_configs`, all series must still be unique; otherwise, the scrape will be rejected for duplicate time series.

To drop the series, we'll match the metric name and use the `drop` action. Since we're using the Prometheus Operator via kube-prometheus, we need to apply the `metric_relabel_configs` settings to the `ServiceMonitor` object that corresponds to the job that `kubernetes_feature_enabled` comes from (the `apiserver` and `kubelet` jobs). This is how it's done in the Helm chart's values:

```
kubeApiServer:
  serviceMonitor:
```

```
    metricRelabelings:
    # Drop excessively noisy apiserver buckets.
    - action: drop
      regex: apiserver_request_duration_seconds_
bucket;(0.15|0.2|0.3|0.35|0.4|0.45|0.6|0.7|0.8|0.9|1.25|1.5|1.75|2|3|
3.5|4|4.5|6|7|8|9|15|25|40|50)
      sourceLabels:
        - __name__
        - le
    - action: drop
      regex: kubernetes_feature_enabled
      sourceLabels: [ __name__ ]
kubelet:
  serviceMonitor:
    metricRelabelings:
      - action: drop
        regex: kubernetes_feature_enabled
        sourceLabels: [ __name__ ]
```

The additional relabeling in kubeApiServer for apiserver_request_duration_seconds_ bucket is a default value, so we include it so as not to overwrite it when adding the relabeling for kubernetes_feature_enabled.

After applying it, we should see that metric disappear in the web UI:

```
$ helm upgrade --namespace prometheus \
    --version 47.0.0 \
    --values ch7/values.yaml \
    mastering-prometheus \
    prometheus-community/kube-prometheus-stack
```

It will take a while for the head cardinality stats page to update since the metrics still exist in the head block (unless you delete them) – so don't be thrown off by that.

Now that we know how to retroactively identify high cardinality labels, let's look at how to prevent it from occurring in the first place.

Using limits

A somewhat recent addition to Prometheus is the ability to set limits on scrape jobs to help protect against cardinality explosions. Rather than retroactively dropping metrics or labels that you find to be high cardinality, you can prevent them from ever being ingested into your Prometheus instances. Additionally, this added layer of safety can enable you to empower users to take more control over sending their metrics to your Prometheus instance without fear of negatively impacting the Prometheus instance.

There are four different types of limits:

- **Sample limits**: Limit the number of samples (that is, series) that a single scrape can contain
- **Label limits**: Limit the total number of labels that can be attached to a sample
- **Label value length limits**: Limit how long any given value for a label can be based on the number of characters (including whitespace)
- **Label name length limits**: Limit how long the name of a label can be, also based on the number of characters in the string

Limits can be defined both globally and per scrape job, so let's take a look at how to do both.

Global limits are defined under the top-level `global` key in the Prometheus configuration file, like so:

```
global:
  sample_limit: 2000
  label_limit: 20
  label_value_length_limit: 50
  label_name_length_limit: 20
```

When set under the global key, these limits will apply to *all* scrape jobs unless explicitly overridden in a scrape job. How do you set the limits in an individual scrape job? I'm glad you asked.

We can use the same keys for this, but under an individual scrape job, we must set the limits:

```
scrape_configs:
  - job_name: exampleJob
    sample_limit: 10000
    label_limit: 15
    label_value_length_limit: 20
    label_name_length_limit: 12

    kubernetes_sd_configs: [ . . . ]
```

All limits are evaluated *after* metric relabeling occurs, so you can still drop unwanted series and/or labels without them counting against the limits. Additionally, if any of the configured limits are breached, then the **entire** scrape is treated as failed and no samples are ingested. Consequently, you should ensure that you have alerting configured for failed scrapes when enabling these limits to not silently be missing metrics due to failed scrapes.

What if too much data coming *in* isn't the issue, but rather too much data going *out*? Queries that read a large number of series and/or samples can be slow and can bog Prometheus down. For that, recording rules can help.

Recording rules

A simple way to improve the performance of your Prometheus queries is through the use of recording rules.

When performing complex queries over long time ranges, your query execution times can get quite long. If you're putting these queries in dashboards in Grafana – or otherwise frequently using them – it can negatively impact user experience. Fortunately, Prometheus enables the pre-computation of these expensive queries via recording rules.

Recording rules differ from alerting rules because they produce *new* time series based on the results of the PromQL expression they evaluate. They can be defined alongside alerting rules, including within the same rule groups.

Recording rule conventions

Since recording rules produce new time series, they need to have unique names. Additionally, it should be made clear that a metric is created by a recording rule. Prometheus does not enforce naming restrictions for recording rules differently than naming restrictions for normal metrics. However, there is a well-established convention for the naming of recording rules using colons (:) as a delimiter.

The pattern is as follows:

```
level:metric:operations
```

Here, `level` is the labels that `metric` is aggregated by and `operations` are the PromQL query functions applied to `metric`. There are a few additional conventions within this naming scheme:

- If multiple labels are used for `level` (or multiple `operations` are applied), they are all listed in their applicable part of the name and separated by an underscore (_).

  - When aggregating using `without` instead of `by`, it generally doesn't make sense to include every label that will be present in the level, so only include the most relevant labels.

- The order of `operations` is newest first (for example, a `sum` value of a 5m `rate` would be `:sum_rate5m`).

- When applying `rate` or `irate` functions to a counter (that is, a metric with a `_total` suffix), strip that suffix in the `metric` name of your recording rule.

Here's an example of how kube-prometheus uses a recording rule to perform potentially expensive aggregations to produce a useful metric from a recording rule named node_namespace_pod_container:container_cpu_usage_seconds_total:sum_irate:

```
groups:
- name: k8s.rules
  rules:
```

```
    - expr: |-
        sum by (cluster, namespace, pod, container) (
          irate(container_cpu_usage_seconds_total{job="kubelet",
  metrics_path="/metrics/cadvisor", image!=""}[5m])
          ) * on (cluster, namespace, pod) group_left(node) topk by
  (cluster, namespace, pod) (
            1, max by(cluster, namespace, pod, node) (kube_pod_
  info{node!=""})
          )
        record: node_namespace_pod_container:container_cpu_usage_
  seconds_total:sum_irate
```

Using recording rules liberally can significantly improve the user experience when users are consuming data from your Prometheus instances. Nevertheless, be careful to ensure that you use recording rules appropriately. The more recording rules you add, the more that the Prometheus TSDB is constantly being queried in the background. This is likely to add to CPU and memory utilization as data points are constantly being loaded into memory. In some Prometheus instances I've seen, over a quarter of the system's RAM was taken up by Prometheus's rule manager evaluating recording rules!

Chaining recording rules

Be wary of chaining recording rules by depending on the results of one recording rule within another recording rule within the same group. Rules within a rule group are evaluated sequentially, so you must order them in the proper evaluation order if you want to have accurate and current results. If possible, avoid chaining rules together like this at all to prevent potential confusion.

In addition to the actual data going into and out of Prometheus, there is another – less discussed – problem that can arise regarding how the data gets into Prometheus – specifically, how Prometheus schedules and executes its scrape jobs, and how slight timing differences can lead to **scrape jitter**.

Scrape jitter

Scrape jitter is the most common cause of oversized TSDB blocks that I have observed. Recall from *Chapter 3* how – from the third scrape onwards – timestamp values are stored in the TSDB as they only store the **delta of the delta** of the sample timestamp. So long as this delta of the delta is 0, the TSDB's compaction process can save a lot of space by effectively dropping the timestamp value from stored samples that all occur at a consistent delta. With millions of samples, this can add up to gigabytes of storage space in every TSDB block. However, when the delta of the delta is *not* consistent, this is referred to as **scrape jitter**.

Scrape jitter is a way to say that scrapes do not occur at consistent intervals. In Prometheus, this often means that they are off by just a few milliseconds. By default, Prometheus will automatically adjust timestamps that are within a 2ms tolerance.

> **Configuring timestamp adjustments**
>
> Whether or not timestamps are adjusted can be configured via the `--scrape.adjust-timestamps` command-line flag when starting Prometheus. Similarly, the amount of tolerance can be configured via the `--scrape.timestamp-tolerance` flag.

Identifying scrape jitter can be done by querying time series over a time range and evaluating the delta between timestamps. One of the Prometheus maintainers, Julien Pivotto, created a handy tool to do this automatically with a command-line tool called `oy-scrape-jitter`.

The tool can be downloaded from its GitHub project page at `https://github.com/ollydev/oy-toolkit/releases`. To run it, point it at a Prometheus URL and have it log the results, like so:

```
$ ./oy-scrape-jitter \
    --prometheus.url=http://localhost:9090 \
    --log.results-only
```

```
level=info aligned_targets=2 unaligned_targets=22 max_ms=30
```

As you can see from my output, I have several targets that are not aligned with the default 2ms tolerance. So long as there is not some underlying issue causing scrapes to be out of alignment, I can adjust my tolerance to ensure I achieve better TSDB compression.

The `--scrape.timestamp-tolerance` flag can be – at most – up to 1% of the scrape interval. Since my default scrape interval is 30s, I can only set the flag up to 300ms. However, my maximum misalignment is 30ms, so I'll start there. To adjust tolerance with our Helm chart, we can add the following value:

```
prometheus:
  prometheusSpec:
    additionalArgs:
      - name: "scrape.timestamp-tolerance"
        value: "30ms"
```

Then, we must apply the updated Helm chart:

```
helm upgrade --namespace prometheus \
    --version 47.0.0 \
    --values ch7/values.yaml \
    mastering-prometheus \
    prometheus-community/kube-prometheus-stack
```

Shortly after updating the value of the flag, we can re-run the `oy-scrape-jitter` tool to confirm that samples are better aligned:

```
$ ./oy-scrape-jitter \
    --prometheus.url=http://localhost:9090 \
    --log.results-only

level=info aligned_targets=23 unaligned_targets=1 max_ms=34
```

Don't waste too much energy trying to ensure that all of your scrapes are always perfectly aligned, though. There are bound to be circumstances in which resource contention leads to a scrape being started slightly later than it should have started, but that's alright. So long as the majority of the scrapes are aligned, you shouldn't need to worry about scrape jitter. Still, I recommend checking it on a somewhat regular cadence, or at least whenever your Prometheus instance's disk begins to fill up.

Using pprof

On occasion, you may wish to have insight into how Prometheus is performing under the hood at a much lower level. For example, what specifically in the Prometheus code is causing my memory usage to be what it is? Or what is using so much CPU? Since Prometheus is written in Go, it can leverage Go's native **pprof** system.

pprof is a tool that came out of Google (`https://github.com/google/pprof`) that is used to produce, visualize, and analyze profiling data from applications. Prometheus exposes pprof endpoints via HTTP under the `/debug/pprof` path (for example, `http://localhost:9090/debug/pprof`):

/debug/pprof/

Set debug=1 as a query parameter to export in legacy text format

Types of profiles available:
Count Profile
23494 allocs
0   block
0   cmdline
234   goroutine
23494 heap
0   mutex
0   profile
10   threadcreate
0   trace
full goroutine stack dump

Profile Descriptions:

- allocs: A sampling of all past memory allocations
- block: Stack traces that led to blocking on synchronization primitives
- cmdline: The command line invocation of the current program
- goroutine: Stack traces of all current goroutines. Use debug=2 as a query parameter to export in the same format as an unrecovered panic.
- heap: A sampling of memory allocations of live objects. You can specify the gc GET parameter to run GC before taking the heap sample.
- mutex: Stack traces of holders of contended mutexes
- profile: CPU profile. You can specify the duration in the seconds GET parameter. After you get the profile file, use the go tool pprof command to investigate the profile.
- threadcreate: Stack traces that led to the creation of new OS threads
- trace: A trace of execution of the current program. You can specify the duration in the seconds GET parameter. After you get the trace file, use the go tool trace command to investigate the trace.

Figure 7.2 – The /debug/pprof page on a Prometheus server

Although you can view all of these endpoints, such as /debug/pprof/heap and /debug/pprof/goroutine, in the browser, they are best used with the go command-line program. Go has the built-in ability to retrieve and visualize pprof data. For example, you can retrieve heap data like so:

```
$ go tool pprof http://localhost:9090/debug/pprof/heap
```

This produces an output similar to the following:

```
Fetching profile over HTTP from http://localhost:9090/debug/pprof/heap
Saved profile in /home/whegedus/pprof/pprof.prometheus.alloc_objects.
alloc_space.inuse_objects.inuse_space.002.pb.gz
File: prometheus
Type: inuse_space
Time: Sep 7, 2023 at 10:17pm (EDT)
```

This drops you into an interactive shell to manipulate and visualize the data:

```
Entering interactive mode (type "help" for commands, "o" for options)
(pprof)
```

From this shell, you can use a variety of commands (all viewable by typing `help` into the shell). The two most helpful – in my experience – are `top` and `web`.

The output of `top` will show you the top 10 items relative to what `pprof` endpoint you are looking at. In our example of looking at the heap endpoint, we'll see the top 10 memory users:

```
(pprof) top
Showing nodes accounting for 82.47MB, 67.76% of 121.71MB total
Dropped 137 nodes (cum <= 0.61MB)
Showing top 10 nodes out of 123
      flat  flat%   sum%        cum   cum%
   17.01MB 13.98% 13.98%    17.01MB 13.98%  github.com/prometheus/
prometheus/model/labels.(*Builder).Labels
   11.44MB  9.40% 23.37%    11.44MB  9.40%  github.com/prometheus/
prometheus/scrape.(*scrapeCache).addRef
   10.89MB  8.94% 32.32%    10.89MB  8.94%  github.com/prometheus/
prometheus/scrape.newScrapePool.func1
       9MB  7.40% 39.71%        9MB  7.40%  github.com/prometheus/
prometheus/tsdb/chunkenc.NewXORChunk
    7.74MB  6.36% 46.08%     7.74MB  6.36%  github.com/golang/snappy.
Encode
    7.71MB  6.33% 52.41%     7.71MB  6.33%  github.com/prometheus/
prometheus/tsdb/index.(*MemPostings).addFor
       6MB  4.93% 57.34%        9MB  7.40%  github.com/prometheus/
prometheus/tsdb.newMemSeries
    5.19MB  4.26% 61.60%     5.19MB  4.26%  github.com/prometheus/
prometheus/scrape.(*scrapeCache).trackStaleness
       4MB  3.29% 64.89%        4MB  3.29%  bufio.NewWriterSize
(inline)
    3.50MB  2.88% 67.76%     7.50MB  6.16%  github.com/prometheus/
prometheus/tsdb.(*memSeries).mmapCurrentHeadChunk
```

As we'd expect, the most amount of memory is used by keeping track of labels on time series. However, this view is not the most useful. I'm a more visual person and would love to see the call stack that leads to memory allocations on the heap. Using the `web` command can help with that!

To use the `web` command (or any other `pprof` command that renders a graph), we must install Graphviz from `https://www.graphviz.org/download/`. Once installed, we can run the `web` command to open `pprof` data in a web browser:

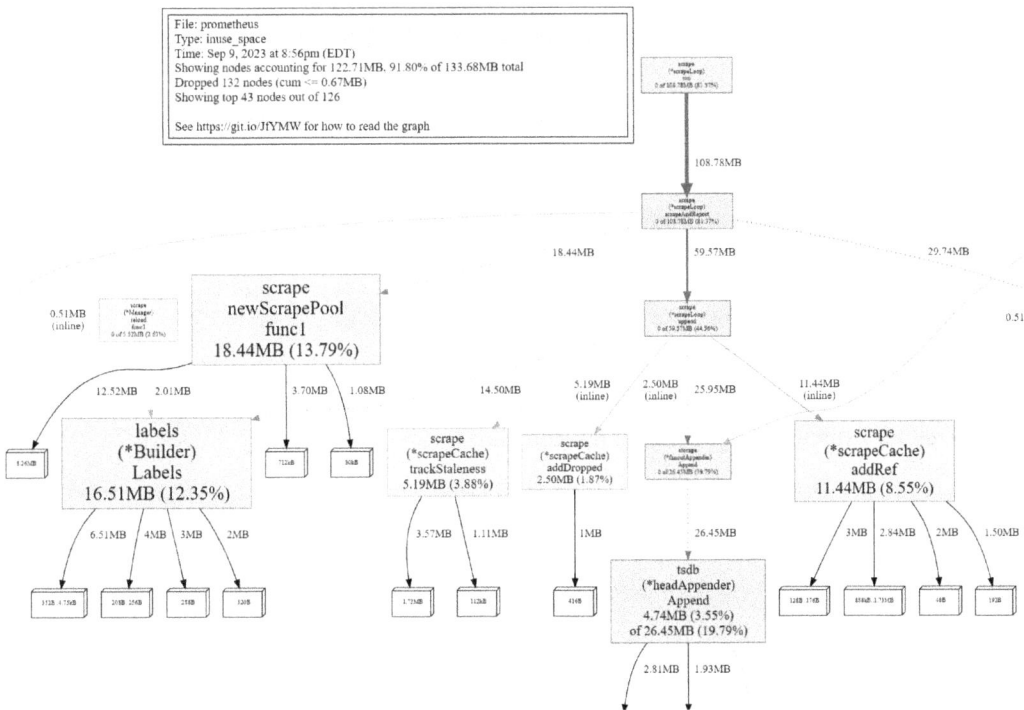

Figure 7.3 – pprof's web output

This makes it a lot clearer to see where memory is coming from! Keep in mind, though, that this only accounts for memory on the *heap*. A discussion of Go's memory model is beyond this book, but suffice it to say that the **heap** is long-lived memory, whereas memory allocated on the **stack** is short-lived and scoped to a function. The important bit is that you should not expect the memory reported by pprof's heap endpoint to match up with the overall memory usage you see Prometheus using with 100% accuracy.

Using promtool for pprof data

You don't have to be a Go developer to be able to leverage pprof and get data from Prometheus's pprof endpoints. The Prometheus developers integrated functionality into the promtool binary – which ships with Prometheus by default – to allow you to capture pprof data and save it to a file without even using the go command-line tool.

Using promtool debug pprof, you can capture pprof data from all available endpoints and store it in a compressed .tar file for later analysis:

```
$ promtool debug pprof http://localhost:9090
```

Running this will produce output that should look similar to the following:

```
collecting: http://localhost:9090/debug/pprof/threadcreate
collecting: http://localhost:9090/debug/pprof/profile?seconds=30
collecting: http://localhost:9090/debug/pprof/block
collecting: http://localhost:9090/debug/pprof/goroutine
collecting: http://localhost:9090/debug/pprof/heap
collecting: http://localhost:9090/debug/pprof/mutex
collecting: http://localhost:9090/debug/pprof/trace?seconds=30
Compiling debug information complete, all files written in "debug.tar.
gz".
```

You can then copy this file off of the Prometheus server and analyze it on your local machine, which can be incredibly helpful both for personal use and filing bug reports.

Alternatively, you can use `promtool debug all`, which will also gather the metrics from the Prometheus server itself and include them in the output.

> **Best practices**
>
> It's become a standard practice of most members of my team to run `promtool debug all` before restarting Prometheus when they need to due to unexplained performance problems. This ensures that they capture data that can help them pinpoint a root cause later.

There is a myriad of uses for `pprof` and what we've discussed just scratches the surfaces. For further information, I recommend reading through the references concerning `pprof` that are provided at the end of this chapter.

Query logging and limits

Depending on your environment and use case, you may end up in a situation in which you are experiencing performance problems due to expensive queries being run or a high number of queries. Prometheus provides built-in ways to log completed queries, maintain a log of active queries, and set limits to prevent overly broad queries from pulling in too much data. Using these, we can both determine the source of issues and put safeguards in place to prevent them from happening.

Query logging

Query logging can be configured via the Prometheus configuration file under the `global` key with the `query_log_file` setting. This setting accepts a file path to which Prometheus will write all queries that it completes:

```
global:
    query_log_file: /var/log/prometheus/my_prometheus_queries.log
```

Each entry in the query log file includes helpful information such as the timestamp of the query, the query itself, and other useful statistics about the query. An example log line might look like this:

```
{
    "params": {
        "end": "2023-09-14T02:49:53.242Z",
        "query": "max_over_time(reloader_last_reload_
successful{namespace=~\".+\"}[5m]) == 0",
        "start": "2023-09-14T02:49:53.242Z",
        "step": 0
    },
    "ruleGroup": {
        "file": "/etc/prometheus/rules/rules_file.yaml",
        "name": "config-reloaders"
    },
    "spanID": "0000000000000000",
    "stats": {
        "timings": {
            "evalTotalTime": 0.000173986,
            "resultSortTime": 0,
            "queryPreparationTime": 0.000103703,
            "innerEvalTime": 0.000063172,
            "execQueueTime": 0.000017709,
            "execTotalTime": 0.000197365
        },
        "samples": {
            "totalQueryableSamples": 0,
            "peakSamples": 2
        }
    },
    "ts": "2023-09-14T02:49:53.243Z"
}
```

In this example, the query that was executed was from an alerting rule and we can also see what specific file it came from. Since the logs are in JSON format, you can use command-line tools such as jq to filter and easily manipulate the logs when you're searching for issues or patterns.

Notably, this query log only contains queries once they have finished executing. Prometheus's TSDB has a separate file it maintains for tracking queries that are currently being executed. This file is automatically created and managed by Prometheus without it having to be enabled. It is stored under the Prometheus TSDB's data directory in a file named queries.active.

If your Prometheus instance restarts, the queries.active file will be cleared out, but Prometheus will also print out a log line stating which queries were still running when Prometheus was shut down.

Query limits

While being able to see what queries your Prometheus ran (or is running) is certainly helpful for identifying issues, that is much more *reactive* than *proactive*. How can we prevent queries from getting out of control in the first place? You're in luck because Prometheus has settings for that!

Prometheus allows you to tune both how many queries can be executed simultaneously *and* how broad those queries are allowed to be. Both settings can be tweaked via command-line flags that are passed to the Prometheus process at startup and are named `--query.max-concurrency` and `--query.max-samples`, respectively.

The `--query.max-concurrency` flag determines how many queries Prometheus can be executing simultaneously before it starts to reject new requests. It defaults to `20`, which is going to be pretty low for larger environments.

The `--query.max-samples` flag, on the other hand, limits how many samples any individual query can attempt to retrieve. It defaults to a comparatively high value of `50,000,000`. Each sample represents a data point, so this flag controls how many individual data points can be touched by a query.

Note that this does not necessarily mean how many samples are *returned* by a query but how many samples must be loaded to process the query. For example, an instant query taking the `rate()` function of a metric over 5 minutes with a 30s scrape interval means that 10 samples are loaded into memory, even though only one data point is returned.

Tuning garbage collection

The Go **garbage collector (GC)** is surprisingly simple if you're coming from a background of other garbage-collected languages such as Java, which has a seemingly unlimited number of ways to tune garbage collection. There are a very limited number of ways to tune the GC, so we'll focus on the two primary ways: the `GOGC` and `GOMEMLIMIT` environment variables.

Until recently (Go 1.19), the `GOGC` environment variable was the only supported way to control garbage collection behavior when running a Go program. Effectively, the way that it works is by setting a percentage of how much the live heap size (memory) can increase from the previous garbage collection before the GC kicks in to mark and reclaim memory.

> **Note**
>
> This is an intentional oversimplification as other things, such as the memory size of goroutine stacks and global pointers, can also contribute to the overall memory size and are included in the percentage that memory can be increased by. In practice, these should not factor too much into Prometheus's memory usage.

The default value of 100 means that the heap can double in size from the live data remaining after the previous collection before a new GC cycle is triggered to mark memory for reclamation. This is what that looks like thanks to the Go project's tool for visualizing it:

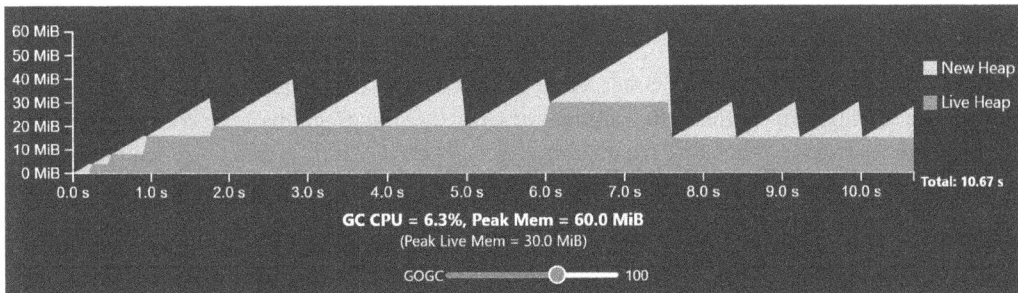

Figure 7.4 – GOGC=100 memory usage over time

As you can see, heap usage is around 20 MiB for a while, so the target heap size is 40 MiB (a 100% increase). Then, when the live heap is 30 MiB, the target is increased to 60 MiB before GC kicks in, and so on.

If we lower GOGC to 50, the heap size will only grow by half before GC kicks in. So, the target for a 20 MiB live heap becomes 30 MiB, the target for a 30 MiB live heap becomes 45 MiB, and so on:

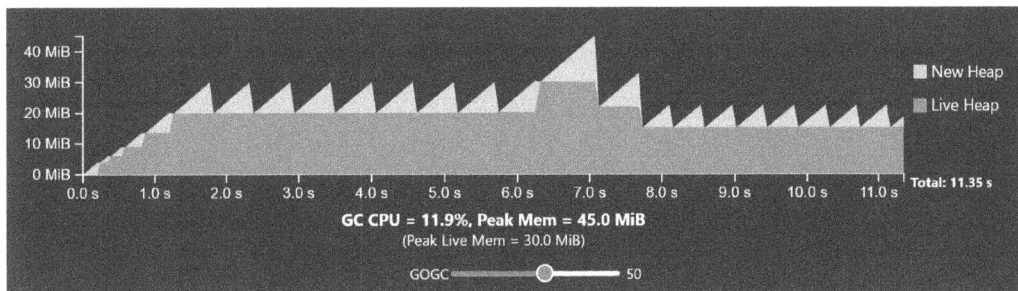

Figure 7.5 – GOGC=50 memory usage over time

Of course, nothing comes without a trade-off. In this case, the trade-off of lowering GOGC is that we trade increased CPU usage for lowered memory usage. In my experience, memory tends to be a more sparse resource – and Prometheus is more memory-hungry than CPU-hungry – so this is a good trade-off.

In practice, I'd recommend tuning GOGC downwards for Prometheus instances with several million or more active time series in the head block. 50 is a good place to start and is what I have seen recommended the most by others. It's also worked well in practice in the Prometheus environments I've administered.

Using GOMEMLIMIT

I mentioned earlier that GOGC was the only way to tune garbage collection until Go 1.19. So, what changed? In Go 1.19, the Go team introduced a new GOMEMLIMIT environment variable for tuning garbage collection.

The GOMEMLIMIT setting does *not* supersede GOGC and can be used in conjunction with it. It functions by providing a soft limit on the maximum amount of memory a Go program can safely use. Whereas GOGC is always relative (and therefore is a moving target), GOMEMLIMIT is static. As memory usage nears the set limit, Go will automatically increase the aggressiveness of its garbage collection.

Consider GOGC with the default percentage value of 100. Without GOMEMLIMIT, using more than 50% of your system's memory puts you in a situation where Go will not trigger a mark phase of its GC until memory usage hits a level that exceeds the total memory on the system (since the garbage will set its next target to double – 100% – of the current memory usage). I've seen this cause issues in Prometheus environments where memory temporarily spikes up due to WAL compaction and – at the same time – someone runs some expensive queries on the Prometheus instance that also increase memory usage temporarily. Its next garbage collection target gets set above the total memory of the server, and it's only a matter of time until the Prometheus process gets hit by the OOM (out-of-memory) killer.

If you choose to set GOMEMLIMIT, the exact setting for it will vary widely based on your environment. However, I would recommend starting with a setting that is roughly 90% of the true total memory available to Prometheus (remember, it's a soft limit, so actual memory usage may exceed the limit). For example, on a 32 GiB server, you could set GOMEMLIMIT to a value of 28 GiB.

Summary

In this chapter, we learned how to make our Prometheus instance more reliable and efficient by applying different tweaks and tunings. We also learned how to debug Prometheus when things go wrong. Now that we have a good handle on running our Prometheus instance, in the next chapter, we'll take an in-depth look at one of the primary sources of metrics in the Prometheus ecosystem: Node Exporter.

Further reading

To learn more about the topics that were covered in this chapter, take a look at the following resources:

- *Scrape Jitter*: https://o11y.tools/oy-scrape-jitter/
- *pprof*:
  - https://pkg.go.dev/runtime/pprof
  - https://jvns.ca/blog/2017/09/24/profiling-go-with-pprof/
  - https://www.robustperception.io/optimising-startup-time-of-prometheus-2-6-0-with-pprof/

- https://www.robustperception.io/analysing-prometheus-memory-usage/

- https://gist.github.com/slok/33dad1d0d0bae07977e6d32bcc010188

- *Garbage collection*:

  - https://go.dev/doc/gc-guide

  - https://weaviate.io/blog/gomemlimit-a-game-changer-for-high-memory-applications

8

Enabling Systems Monitoring with the Node Exporter

Now that we're all experts on running Prometheus itself, it's time to get into what truly sets Prometheus apart: the ecosystem around it. Prometheus has a vibrant community with a multitude of open source projects that extend it and expose data to it. Many Prometheus-related projects are **exporters**.

The term "exporter" in Prometheus refers to any application that runs independently to expose metrics from some other data source that is not exposing Prometheus metrics natively. There are exporters for almost anything you can think of, from MySQL to Minecraft, and a non-exhaustive list can be found on Prometheus's official docs site at `https://prometheus.io/docs/instrumenting/exporters/`. However, in this chapter, we're going to focus on the most popular and common exporter: the Node Exporter.

We're going to cover the following main topics:

- Node Exporter overview
- Default collectors
- The textfile collector
- Troubleshooting the Node Exporter

Let's get started!

Technical requirements

For this chapter, you can connect to the Prometheus cluster we created in *Chapter 2* to follow along with exploring metrics, but the only true requirement for this chapter is optional. It is only needed if you want to experiment with the basics of writing a Prometheus exporter:

- **go**: `https://go.dev/dl/`

Code used in this chapter is available at `https://github.com/PacktPublishing/Mastering-Prometheus`.

> **Note**
>
> This chapter focuses only on the Node Exporter, which is only useful for systems with *NIX kernels (e.g., Linux, FreeBSD, MacOS, etc.). For Windows systems, a similar – but separate – exporter exists called the Windows Exporter (`https://github.com/prometheus-community/windows_exporter`).

Node Exporter overview

The Node Exporter is one of the select few exporters maintained by the official Prometheus project, alongside others such as the Blackbox Exporter and the SNMP Exporter. Its purpose is to expose a variety of machine-level metrics pertaining to resources such as CPU, disk, memory, networking, and more.

One of the things I often say to people asking whether we have some system-level metric in Prometheus is, "If it's in `/proc`, the Node Exporter can get it."

> **What's /proc?**
>
> In Linux systems, a `/proc` directory exists that contains a plethora of information about the state of the machine. The Linux kernel documentation describes it thusly:
>
> *The proc file system acts as an interface to internal data structures in the kernel. It can be used to obtain information about the system and to change certain kernel parameters at runtime (`sysctl`).*

The Node Exporter primarily retrieves data through the `/proc` pseudo-filesystem. There are a few exceptions – such as the hwmon collector, which collects data from `/sys/class/hwmon` – but most collectors leverage the Prometheus `procfs` library (`https://github.com/prometheus/procfs`) to pull data from `/proc` and convert it to Prometheus metrics.

What is in an exporter?

To understand how the Node Exporter works, it's helpful to first step back and look at what makes up an exporter. As we've already discussed, an exporter takes data from some external system and coerces it into Prometheus metrics. Every exporter is comprised of one or more metric definitions, a collector, and a metrics registry.

The metrics registry is what keeps track of all of the Prometheus metrics and their values. Metrics are defined in code and registered to a registry. Then, when the HTTP endpoint serving metrics is hit by Prometheus, the collector is triggered to collect the most up-to-date values of the metrics and return them.

Let's take a look at what this looks like in Go code:

```go
package main

import (
        "log"
        "net/http"

        "github.com/prometheus/client_golang/prometheus"
        "github.com/prometheus/client_golang/prometheus/promhttp"
)

var myMetric = prometheus.NewDesc("exporter_success", "Reflects if the
exporter is working.", []string{}, nil)

type myCollector struct{}

func (c myCollector) Describe(ch chan<- *prometheus.Desc) {
    ch <- myMetric
}

func (c myCollector) Collect(ch chan<- prometheus.Metric) {
        ch <- prometheus.MustNewConstMetric(myMetric, prometheus.
GaugeValue, 1.0)
}

func main() {
        prometheus.MustRegister(&myCollector{})
        http.Handle("/metrics", promhttp.Handler())
        log.Fatal(http.ListenAndServe(":9999", nil))
}
```

Starting at the main() function, we register our collector with the metrics registry. When the collector is registered, the Describe() method on the collector is called behind the scenes to describe what metrics are being added to the registry.

In our Describe() method, we add the description for our metric that we named exporter_ success to the registry. Notably, there is no value yet for our metric; we only told the registry that we plan to add values to it later.

Next, we assign an HTTP handler for the /metrics URL path. Using promhttp.Handler() as the handler will ensure that any registered collectors will run their Collect() method when that URL is hit. Most exporters will have just one collector, and all that is needed to satisfy the requirements of the prometheus.Collector interface is to implement a Collect() method and a Describe() method on your collector.

Finally, we run the HTTP server on port `9999`. Running locally, we could go to `http://localhost:9999/metrics` and see the `exporter_success` metric along with several other metrics that `promhttp` will add for us, including Go runtime statistics and some basic HTTP statistics about the exporter itself.

With that basic example, we're able to see all that is required to run a simple exporter. What makes the Node Exporter unique amongst most other exporters is that it has *dozens* of individual collectors that it runs by default and even more that can be optionally enabled.

> **Doesn't having more collectors make it slow?**
>
> Thankfully, due to Go's concurrency model, the quantity of collectors has a negligible effect on how fast the Node Exporter returns data for a scrape. This is because each of the Node Exporter's collectors is run in its own **goroutine** (this is specific to the Node Exporter as other exporters may not use goroutines in the same way). So, don't worry about trying to remove collectors to speed up the Node Exporter just by limiting the quantity you run.

Since there are so many, let's take a look at some of the default ones that I find to be the most useful.

Default collectors

At the time of writing, there are a whopping 49 different collectors that are enabled by default in the Node Exporter. Many of them are either niche (such as `dmi`) or dependent on your infrastructure (such as filesystem collectors for `xfs` and `zfs`). Rather than go through all of them, we'll take a look at some of the most useful ones to see what info they provide and why you would care about it.

conntrack

The `conntrack` collector exposes metrics related to the Linux kernel's netfilter connection tracking subsystem. This is used to keep track of connections established to your server and can cause issues when the table it uses becomes full.

Two commonly used metrics from this collector are as follows:

Metric	Description
`node_nf_conntrack_entries`	Current number of entries in the connection tracking table
`node_nf_conntrack_entries_limit`	Configured maximum number of entries allowed in the connection tracking table

Table 8.1 – Common conntrack metrics

cpu

The cpu collector is the primary collector you use with the Node Exporter to gain insight into how much of your total CPU you are using. Rather than being emitted as a percentage, as you might see in Linux commands such as top, CPU usage is tracked in seconds spent in each CPU mode.

In order to see what percentage of time a CPU core has spent doing work, you'll need to leverage PromQL to calculate the CPU's non-idle time. Given the node_cpu_seconds_total metric, you can get the CPU utilization percentage using the PromQL rate function:

```
sum without (cpu, mode) (rate(node_cpu_seconds_total{mode!="idle"}
[5m]))
```

In this example, we also remove the cpu and mode labels to give us a sum of all non-idle time. This results in a percentage of non-idle time across all CPUs. If you have one CPU, the maximum value is 1.0. If you have two CPUs, the maximum value will be 2.0, and so on. The reason why the maximum value scales with the number of CPUs is because, for every one second of real-world time, each CPU can complete one second of work.

How can you normalize this CPU percentage to stay between 0 and 100%, then? I generally recommend setting up a recording rule to count how many CPU cores you have on any given server. Then, you can use that to divide results from queries such as the previous one to give a normalized value. The query for that recording rule might look something like this:

```
count without (cpu, mode) (node_cpu_seconds_total{mode="idle"})
```

diskstats

The diskstats collector exposes statistics related to the storage device(s) attached to your server. It provides metrics related to things such as disk I/O and what type of filesystem is on the disk. This can be useful for determining whether your workload is performing a lot of reads or writes to disk.

This collector exposes several metrics, but some of the highlights are the following:

Metric	Description
node_disk_io_time_ seconds_total	Counter showing the amount of time spent doing I/O since the system booted. Can be combined with the rate function to get a percentage of time spent doing I/O per second.
node_disk_read_bytes_ total	Counter for the number of bytes read from the disk since system boot. Combine with rate to see per-second read speeds.
node_disk_written_bytes_ total	Same as node_disk_read_bytes_total but for writes.

Table 8.2 – Common diskstats metrics

filesystem

The filesystem collector is the natural complement to the diskstats collector. Whereas diskstats focuses on the physical (or virtual) disk, the filesystem collector exposes metrics related to the filesystem on the disk.

These metrics include information such as the maximum size of the filesystem and how much space is available in the filesystem. Some important metrics from this collector are as follows:

Metric	Description
node_filesystem_size_bytes	The size of the filesystem in terms of how much space is allocated to it.
node_filesystem_free_bytes	The amount of space remaining in the filesystem. Subtract this metric from node_filesystem_size_bytes to determine how much space is *used* in the filesystem.

Table 8.3 – Common filesystem metrics

loadavg

The loadavg collector exposes the typical 1-minute, 5-minute, and 15-minute load averages commonly used in Linux systems and output by command-line utilities such as uptime.

I mention this collector only to beg you not to use this for your monitoring. The Linux source code in the kernel/sched/loadavg.c file itself sums up my feelings on it:

```
Its a silly number but people think its important.
```

For further reading on the unintuitiveness of Linux load averages and better alternatives, I highly recommend Brendan Gregg's blog post on the matter at https://www.brendangregg.com/blog/2017-08-08/linux-load-averages.html.

For one example of a better alternative, check out the section on the pressure collector in this chapter.

meminfo

The meminfo collector is responsible for exposing all of the Node Exporter metrics with the node_memory_ prefix. These metrics expose information pertaining to RAM on the server such as total memory, memory free, and memory available.

Linux distinguishes between memory *free* and memory *available*, so keep that in mind when crafting alerts or dashboards using these metrics. In Linux, you can think of free memory as completely unused memory that can be immediately used, whereas available memory is currently being used – probably as a buffer or cache – but can be reclaimed if something else needs it.

Some useful metrics include the following:

Metric	Description
`node_memory_MemTotal_bytes`	How much total memory is installed in the system.
`node_memory_MemAvailable_bytes`	How much memory is available to be used currently. This includes both free memory and reclaimable memory.
`node_memory_MemFree_bytes`	How much memory is free for immediate use.

Table 8.4 – Common meminfo metrics

netdev

The `netdev` collector is responsible for providing metrics on your network interfaces. This includes data on how many packets are entering and leaving your network devices, how much data is being transferred in and out, and how many packets are being dropped.

Some common metrics from this collector include the following:

Metric	Description
`node_network_transmit_bytes_total`	How much data has been transmitted by the interface. Use with the `rate` function to see data throughput per second.
`node_network_receive_bytes_total`	Same as the previous metric, but for data coming *into* the server.
`node_network_receive_drop_total`	How many packets coming into the server have been dropped by the interface.
`node_network_transmit_drop_total`	Same as the previous metric, but for dropped packets leaving the server.

Table 8.5 – Common netdev metrics

pressure

As mentioned in the `loadavg` collector section, the `pressure` collector is one of my personal favorites. In Linux kernels that are at least version 4.20, the kernel provides a built-in way to effectively track the three most common bottlenecks in a server: disk I/O, memory, and CPU. This tracking system is called **Pressure Stall Information** – or **PSI** for short – and it measures how often one of these resources caused a delay in work being accomplished by the CPU because that resource was under contention.

Delays for disk and memory resources are divided into two categories: *waiting* and *stalled*. The waiting category is for processes that couldn't make progress because they were waiting on the resource, but the scheduler was still able to find other work to run on the CPU. The stalled category, on the other hand, is for processes that couldn't make progress because they were waiting on the resource and the scheduler couldn't find any other work for the CPU to run. This is generally indicative of a significant problem since you are effectively wasting CPU cycles because you're blocked by resource contention.

The pressure information metrics for the CPU only have a waiting category since, if you're waiting on CPU resources, it's because other things are running on the CPU, and therefore you couldn't possibly be stalled since *something* is running on the CPU and making progress.

Compared to load averages, these provide a much better reflection of whether or not your system is under contention and experiencing issues since a high load average does not always necessarily correlate with actual, impactful problems. Consequently, I recommend alerting on at least the disk and memory pressure metrics from this collector. For memory contention, this may even be able to replace existing, naïve memory alerting you already have such as "X system is using 95% of its memory," since the pressure metrics will be much more indicative of actionable issues. However, the CPU pressure metrics may overlap with existing alerting you already have for high CPU usage and therefore may not be as useful.

The metrics exposed by this collector include the following:

Metric	Description
`node_pressure_memory_stalled_seconds_total`	How much time processes were waiting to run, but couldn't because they were waiting on memory resources and nothing else could make progress either
`node_pressure_io_stalled_seconds_total`	How much time processes were waiting to run, but couldn't because they were waiting on disk resources and nothing else could make progress either

Table 8.6 – Common pressure metrics

Others

We've only looked at a small selection of all of the collectors that are enabled by default in the Node Exporter. As of version 1.6.1 of the Node Exporter, these are all of the other collectors enabled by default on Linux systems:

Collector	Description
arp	Exposes ARP statistics related to layer-2 networking.
bcache	Exposes statistics related to Linux's block layer cache feature.
bonding	Exposes statistics related to network interface bonds.
btrfs	Exposes metrics specific to the btrfs filesystem type.
cpufreq	Exposes statistics related to CPU frequencies.
dmi	Exposes information about the Desktop Management Interface.
edac	Exposes statistics about Linux's error detection and correction system. Useful for detecting issues with physical memory sticks (DIMMs).
entropy	Exposes entropy statistics (used for randomness and cryptography).
fibrechannel	Exposes statistics related to Fibre Channel interfaces (a protocol used for high-speed data transfers, typically between storage devices and servers).
filefd	Exposes statistics related to file descriptors on Linux systems.
hwmon	Exposes statistics on supported physical hardware connected to the server.
infiniband	Exposes statistics related to InfiniBand interfaces. InfiniBand is another high-throughput, low-latency protocol like Fibre Channel.
ipvs	Exposes statistics related to **IP Virtual Server** (**IPVS**), which is a layer-4 load-balancing feature built into the Linux kernel.
mdadm	Exposes statistics related to RAID arrays.
netclass	Exposes information on network interfaces.
nfs	Exposes statistics related to **Network File System** (**NFS**) shares attached to the server.
nfsd	Exposes statistics related to NFS shares being served from the server.
nvme	Exposes information on NVMe storage devices attached to the server.
os	Exposes information about the operating system installed on the server.
powersupplyclass	Exposes statistics related to the physical power supply installed on the server.
rapl	Exposes statistics related to Linux's power-capping features.
schedstat	Exposes statistics related to the Linux scheduler's performance.
selinux	Exposes statistics related to SELinux if it is enabled.
sockstat	Exposes statistics on Linux sockets.
softnet	Exposes statistics related to softnet, which is a specific type of softirq used by the Linux kernel for network processing.

Collector	Description
stat	Exposes various system statistics such as boot time, interrupts, and context switches.
tapestats	Exposes statistics related to tape-based storage devices attached to the server.
textfile	Provides a way to read metrics from files on the server and include them in the metrics output. We'll cover this more later in the chapter.
thermal_zone	Exposes statistics related to temperatures in the server.
time	Exposes current system time on the server.
timex	Exposes statistics related to time synchronization on the server (e.g., chronyd, NTP, etc.).
udp_queues	Exposes transmit and receive queue lengths for UDP packets in the system.
uname	Exposes basic system information such as kernel version and the server's hostname.
vmstat	Exposes virtual memory statistics.
xfs	Exposes statistics related to the XFS filesystem.

Table 8.7 – Other available Node Exporter collectors

The list of collectors is ever-expanding, so be sure to check the latest Node Exporter releases to see the latest list.

The textfile collector

The textfile collector is a hidden gem in the Node Exporter. This single collector adds tremendous versatility to what you can accomplish with a Prometheus monitoring stack.

Using the textfile collector, you can read Prometheus-formatted metrics from files on the server and include them in the output of the Node Exporter's /metrics scrape endpoint.

Being able to read metrics from a file opens up a whole new world of monitoring short-lived processes such as batch or cron jobs, where it isn't possible or doesn't make sense to expose metrics on an HTTP endpoint for Prometheus to scrape. For example, at my company, we leverage the textfile collector to expose metrics related to the last time a server executed its scheduled synchronization with our configuration management system.

The textfile collector is enabled by default but requires additional configuration to actually work. For the collector to work, it must know where it should look for files with metrics in them. This is specified via the --collector.textfile.directory flag.

The flag is not repeatable, but it does support shell-style file globs such as /var/lib/prometheus/*/*, which would discover files with a .prom file extension in /var/lib/prometheus/cron_jobs/job1/ and /var/lib/prometheus/cron_jobs/job2/, but not /var/lib/prometheus/my_app/.

For metrics to be discovered, they *must* have a .prom file extension, such as /var/lib/prometheus/cron_jobs/job1/metrics.prom. This is not configurable and is hardcoded into the collector.

Additionally, metrics must be in the standard Prometheus text-based format described at https://prometheus.io/docs/instrumenting/exposition_formats/#text-based-format. It's the same format that you see your metrics exposed in over HTTP, and looks like this:

```
# HELP my_job_last_run_time_seconds The last time that my_job
completed a run.
# TYPE my_job_last_run_time_seconds gauge
my_job_last_run_time_seconds 1572210000

# HELP my_job_last_run_success Whether the last run of my_job was
successful.
# TYPE my_job_last_run_success gauge
my_job_last_run_success 1
```

As versatile as it is, be wary of getting carried away with overusing the textfile collector. It introduces added complexity and the possibility for silent failures where a file is no longer being updated, because the process that is supposed to write to it is broken, but the most recent metrics written to the file still indicate that it's working correctly.

To protect against these silent failures, the Node Exporter exposes a metric called node_textfile_mtime_seconds, which shows the last time a metrics file read by the textfile collector was modified. If you know how often a file should be updated, I highly recommend adding alerts for times when the modified time exceeds that interval.

The textfile collector is ultimately still a part of the Node Exporter, so try to avoid getting carried away with adding metrics using it for things that are unrelated to the machine it is running on, such as service-focused metrics. You'd likely be better off leveraging the Prometheus Pushgateway for that. The Pushgateway is not covered in this book, but additional documentation on it can be found on its GitHub project page at https://github.com/prometheus/pushgateway.

Finally, for some examples of good usages of the textfile collector to augment system-level metrics, I'd recommend checking out the Prometheus community's collection of scripts for ideas at https://github.com/prometheus-community/node-exporter-textfile-collector-scripts/.

Troubleshooting the Node Exporter

I would be remiss if I gave you the impression that the Node Exporter just magically works 100% of the time. Undoubtedly, you'll experience issues where Node Exporter scrapes begin experiencing issues such as slow scrapes or even timeouts. Thankfully, the Node Exporter provides us with some per-collector metrics to help pinpoint where the issue lies.

The `node_scrape_collector_success` metric returns whether or not running an individual collector was successful. But wait – before you go putting alerts in for any time any `node_scrape_collector_success` time series returns a 0, remember that not all of the collectors that are enabled by default are expected to apply to your system. For example, I seriously doubt your server has both InfiniBand and Fibre Channel connections (most likely you have neither), so something's always going to be marked as failing.

Instead, the metric I tend to look at the most for Node Exporter troubleshooting is `node_scrape_collector_duration_seconds`. This metric exposes how long it took to run each collector. In my experience, it's generally just one collector that is causing the slowness.

As an anecdote, I was once troubleshooting an issue with the Node Exporter where scrapes were frequently timing out and causing gaps in graphs. Using the `node_scrape_collector_duration_seconds` metric, we were able to quickly pinpoint that the issue was that the `processes` collector had been enabled on the Node Exporter instance running on the server. This collector is disabled by default, and it was accounting for all of the slowness since it was competing with an application for a lock on a specific system resource. The time it took to finally receive the lock was intermittently longer than the scrape timeout, so we'd lose out on all data from the Node Exporter.

Summary

In this chapter, we learned all about Prometheus's most popular exporter, the Node Exporter. We went over the basics of what an exporter is and what is involved in creating one. Then, we dove headfirst into the dozens of collectors that the Node Exporter enables by default. Finally, we looked at how to use the `textfile` collector and how to troubleshoot issues with the Node Exporter.

In our next chapter, we're going to be stepping out of the realm of "vanilla" Prometheus and begin looking at how we can extend and augment Prometheus through the use of other open source projects. To begin, we'll see how projects such as VictoriaMetrics and Grafana Mimir can function as remote storage for Prometheus metrics.

Further reading

To learn more about the topics that were covered in this chapter, take a look at the following resources:

- *The /proc filesystem*: `https://www.kernel.org/doc/html/latest/filesystems/proc.html`

- *Understanding and Building Exporters*: `https://training.promlabs.com/training/understanding-and-building-exporters`

- *PSI - Pressure Stall Information*: `https://www.kernel.org/doc/html/latest/accounting/psi.html`

- *Awesome Prometheus alerts for Node Exporter*: `https://samber.github.io/awesome-prometheus-alerts/rules#host-and-hardware`

Part 3: Extending Prometheus

In our final part, we dive into the most advanced topics yet. We've learned the fundamentals of Prometheus itself and how to scale Prometheus using just Prometheus, but now we delve into the territory of extending Prometheus by introducing additional technologies to augment it.

First, we will look at how to extend Prometheus both centrally – through remote storage systems such as Grafana Mimir and VictoriaMetrics – and in a more distributed manner – through the use of Thanos. Next, we'll look at how various tools can make the management of Prometheus at scale easier by simplifying the management of Prometheus rules via Jsonnet, automatic validation and testing of Prometheus configurations and rules in CI pipelines, and generating **Service Level Objectives (SLOs)** from Prometheus metrics that follow best practices. Finally, we'll begin looking to the future by exploring the OpenTelemetry project – how it integrates with Prometheus for metrics and more – and what comes next after this book to continue your observability journey beyond Prometheus.

This part has the following chapters:

- *Chapter 9, Utilizing Remote Storage Systems with Prometheus*
- *Chapter 10, Extending Prometheus Globally with Thanos*
- *Chapter 11, Jsonnet and Monitoring Mixins*
- *Chapter 12, Utilizing Continuous Integration (CI) Pipelines with Prometheus*
- *Chapter 13, Defining and Alerting on SLOs*
- *Chapter 14, Integrating Prometheus with OpenTelemetry*
- *Chapter 15, Beyond Prometheus*

Utilizing Remote Storage Systems with Prometheus

## 9

On its own, Prometheus can take you pretty far in establishing your monitoring stack and making systems more observable with metrics. Nevertheless, Prometheus is intentionally limited in its feature set and functionality; this is what helps keep it simple and flexible. With Prometheus's huge community support, several other open source projects have arisen to augment and extend Prometheus' base features. Throughout the remainder of this book, we'll be focusing on these projects and how they can help you make the most of your Prometheus environment.

To begin with, we're going to look at Prometheus' remote storage system, which contains methods for both remotely reading and remotely writing Prometheus data. We'll be looking at two popular open source projects that leverage these APIs, including how to deploy them and hook them up to your Prometheus instances.

In this chapter, we're going to cover the following main topics:

- Understanding remote write and remote read
- Using VictoriaMetrics
- Using Grafana Mimir

Let's get started!

Technical requirements

For this chapter, we'll continue building off of the Prometheus environment we deployed to Kubernetes in *Chapter 2*. Consequently, you'll need the two following tools installed:

- **kubectl**: `https://kubernetes.io/docs/tasks/tools/#kubectl`
- **helm**: `https://helm.sh/docs/intro/install/`

The code used in this chapter is available at `https://github.com/PacktPublishing/Mastering-Prometheus`.

Understanding remote write and remote read

Prometheus is purposely limited in the functionality it endeavors to implement. Notably, its storage is tied to a local filesystem, and there is no clustering or replication support for Prometheus instances. When running Prometheus at a sufficient scale, you will inevitably encounter the desire to scale Prometheus beyond these limitations. Rather than constrain you to only using Prometheus, Prometheus has built-in functionality to integrate with other metrics storage systems that are more featureful (but also more complex).

Prometheus does this by exposing two remote storage APIs—one for reading data from a storage system and one for writing data to it. Both APIs are exposed via HTTP, use **protocol buffers** (**protobufs**) for communication, and use "snappy" compression for both requests and responses.

Remote write is certainly the more popular of the two since it involves sending data to a larger metrics system, which you're more likely to query directly at that point. However, remote read still has its uses.

Remote read

Starting off with remote read, Prometheus enables you to connect to external storage systems that can be queried alongside the Prometheus metrics stored in the local Prometheus instance. This could be another Prometheus instance, an external metrics storage system such as Mimir or InfluxDB, or even a custom adapter for a non-metrics data source. For example, you could build an adapter that receives a remote read request, queries data from a SQL database, and returns data formatted as a Prometheus metric.

Remote read is flexible, so you can do some neat things with it if you so choose. However, in practice, most people are not connecting external storage systems to their Prometheus instances to query them from Prometheus. Rather, they're connecting Prometheus to another storage system so that they can query that storage system. Part of the reason for this is the limited control that you have over when queries will hit your configured remote read endpoint. If the system you're connecting to has issues, you don't want your queries (especially your alert queries!) to start failing.

Even so, remote read does have some more common use cases. Thanos, for example, uses Prometheus's remote read API to query data via the Thanos Sidecar component. This is a slightly different scenario in which you're retrieving data from Prometheus via remote read rather than Prometheus itself retrieving data from another system via remote read. That's one of the neat things about both remote read and remote write—a Prometheus instance is able to be both a client and a server for both APIs.

Regardless, remote read is an advanced feature that has not seen much development since its initial release. You are unlikely to need to think about it much (or at all) in practice. Nevertheless, it's good to know that the feature exists, as you may end up having a use case for it one day!

Remote write

Remote write is the far more popular and widely used Prometheus remote storage API. Generally, it is used to write data from Prometheus to an external storage system. This includes sending data to destinations such as AWS's Amazon Managed Service for Prometheus, Cortex, VictoriaMetrics, Grafana Mimir, and many more.

It can also be used to write data to Prometheus, provided that you start Prometheus with the `--web.enable-remote-write-receiver` flag passed to the process. This is used by applications such as Grafana Loki—a logging platform—that can evaluate recording rules based on queries of log data and send the resulting metrics to a Prometheus instance.

Similar to federation, remote write can send all data from Prometheus or only a subset of metrics. Additionally, like federation, the subset of metrics being sent is controlled using relabel configs to select which metrics to keep or drop.

> **Important note**
>
> You may have heard that in version 2.47.0 of Prometheus, a new feature was added to Prometheus to allow Prometheus to receive data via the **OpenTelemetry Line Protocol** (**OTLP**). This is explicitly distinct from remote writing data to another Prometheus instance and will be covered in *Chapter 14*, where we will discuss integrating Prometheus with OpenTelemetry.

Prometheus agent

Remote write is such a popular feature that, in version 2.32.0, Prometheus added support for an "agent" mode that is purely focused on sending metric data over the remote write protocol. You can enable this feature by passing the `--enable-feature=agent` flag to the Prometheus process when starting it.

When running Prometheus in this agent mode, the querying, alerting, and local storage of data are all disabled. It even has a slightly tweaked WAL implementation in which data are immediately removed from the WAL once it is successfully sent to its remote write destination(s).

> **Important note**
>
> Although the agent's WAL is unique in its implementation, it still builds upon the default WAL setup in Prometheus. This means that it's currently limited to buffering only up to 2 hours of data in the WAL since this is how long a non-agent Prometheus WAL would build up data before flushing a TSDB block to disk. There is an open issue that allows for longer buffering when running in agent mode: `https://github.com/prometheus/prometheus/issues/9607`.

Running Prometheus in agent mode allows you to have a nice balance of both push- and pull-based models of data collection. You get the pull-based pros of things such as an up metric to let you know if data collection is actually working on your expected scrape targets and the push-based pros of being able to easily centralize your metrics for a global view and not worry about maintaining dozens or hundreds of stateful Prometheus instances, each storing their own data.

Be forewarned, though, that using Prometheus in agent mode is still considered experimental at the time of writing this book. It's an advanced use case that you are unlikely to need unless you already have a mature time series storage platform outside of Prometheus (such as one of those we're going to talk about later in this chapter). In my personal experience, I've yet to have needed to use the agent mode, but it is something that my team is trending towards in the future.

Remote write tuning

When using remote write in Prometheus, you are likely to be sending a high quantity of data to whatever your remote write destination is. Consequently, there are quite a few knobs you can twist and levers you can pull to tweak how Prometheus is sending the data to the remote write destination. Aside from the normal settings present in most configuration blocks in Prometheus (such as TLS and auth settings), there are three that we'll focus on for tuning remote write setups: *relabel configs*, *queue configs*, and *metadata configs*.

Write relabel configs

As previously mentioned, remote write enables you to control the subset of metrics you send to your remote write destination. This is similar to how using Prometheus federation allows you to specify which time series you want to federate. However, unlike federation—which requires that you explicitly specify which metrics to federate—remote write will send all of the time series and sample values in Prometheus to the remote write destination by default.

Since all metric data will be sent by default, each remote write configuration block has a `write_relabel_configs` section. This section uses the same syntax as the `relabel_config` sections we looked at in *Chapter 4* and other previous chapters.

By using `write_relabel_configs`, you can restrict the metrics being sent to a remote storage backend based on any labels, including a metric's name (the `__name__` label) or some other arbitrary label.

For example, you may have a team using your Prometheus instance that is running its own remote storage backend and wants to send all metrics related to its service to the backend. They don't care about other teams' metrics in the Prometheus instance, so you could configure a remote write destination that only sends their team's metrics to their backend, such as the following:

```
remote_write:
  - url: https://teamMandalorian.olly.example.com/api/prom/push
    write_relabel_configs:
```

```
    - action: keep
      source_labels: [ "team" ]
      regex: 'mandalore'
```

In this example, only metrics that adhere to both of the following conditions will have their metrics sent to the configured remote write destination:

- Have a `team` label present
- Have a `team` label value of `mandalore`

Keep in mind that these relabelings occur after data ingestion (after even any `metric_relabel_configs`), so any labels that are present on the time series in Prometheus will be available to use in your `write_relabel_configs`.

Obviously, one of the simplest ways to reduce the resources used by remote writing is to reduce the amount of data you're sending, so start here when tuning remote write and drop data you don't need to send if you don't care about them. For example, maybe you only want to send time series from production systems with an `environment="production"` label attached to them.

I would caution you against getting carried away with micro-optimizing for every single metric you're sending, though. Similar to getting carried away with `metric_relabel_configs`, the returns quickly diminish, and your configuration file becomes increasingly difficult to read and maintain.

Queue configs

Besides simply changing the amount of data you're sending via relabeling, tuning how you send the data is the other way in which you can tune remote write. To understand how to tune that via queue configs, let's first look at how Prometheus uses queues and shards to deliver data to remote write destinations.

For each remote write destination that is configured, Prometheus creates a new **queue** in memory. Each queue is divided into a dynamic number of **shards**. The exact number of shards per queue is constantly being re-evaluated by Prometheus in the background based on the rate of incoming data that needs to be sent to the destination. However, we can control both the minimum and maximum number of shards per remote write destination.

All shards are being operated on in parallel in order to send their data to their destination. If any one shard reaches its capacity, Prometheus will pause reading any more data from the WAL into any shards. However, no data loss will occur provided that the shard is able to start making progress again prior to the WAL being truncated (recall from *Chapter 3* that the WAL is truncated every 2 hours after the head block is flushed to disk).

Now that we understand how data is divvied up and operated on, we can competently tune it. When tuning, keep in mind that any changes will also impact the memory usage of the Prometheus server—not just throughput to the destination.

The number of shards that Prometheus uses for remote write should generally not need to be tuned. The default value for the `max_shards` setting is `200`, and for `min_shards`, it is `1`. The value of `max_shards` is not recommended to be tweaked. However, `min_shards` may be raised in order to prevent falling behind when Prometheus starts up and is still calculating how many shards it should scale up to based on your server's load. How high to set it will depend on your configuration for these next settings.

The `capacity` setting determines how many samples a shard can hold and defaults to `2500`. If your Prometheus is ingesting more than 2,500 samples per second (and if you're reading this book, it probably is), then the measly 1 initial shard isn't going to cut it.

> **How many samples am I ingesting per second?**
>
> Since Prometheus exposes metrics about its own TSDB, you can easily get a good estimate of how many samples you're ingesting per second using the following PromQL query. This can inform you how to tune your remote write configuration.
>
> ```
> rate(prometheus_tsdb_head_samples_appended_total[5m])
> ```

In addition to `capacity`, each shard is governed by the `max_samples_per_send` setting, which determines when to send a batch from the shard to the destination. This setting defaults to `500`, so when a shard contains 500 samples, it would trigger a send to the remote write destination. In the event that your data ingestion rate is not filling the shard fast enough, the `batch_send_deadline` setting controls how frequently a batch will be sent if `max_samples_per_send` is not reached and defaults to 5 seconds. Ideally, you should not be hitting this deadline unless your Prometheus is low volume.

Circling back to how to set `min_shards`, I would recommend setting your `min_shards` to at least your sample rate (calculated above) divided by your `capacity`:

```
min_shards = sample_rate / capacity
```

This will ensure that you start off in a place where you're at around 100% usage, and Prometheus can quickly scale up additional shards.

In high-throughput systems, you will likely need to increase your `max_samples_per_send` and `capacity`. An increase in the `max_samples_per_send` setting should generally always be accompanied by an increase in `capacity` to ensure that `capacity` is always at least three times the size of `max_samples_per_send`. However, keep in mind that increases in these settings will also increase memory usage by Prometheus, as more samples are held in memory to be sent. Consequently, it is recommended to tune down the `max_shards` setting to avoid accidentally running out of memory. If you have plenty of headroom, feel free to keep it the same, though.

Finally, what happens when you start hitting errors while sending to your remote write destination? Perhaps you have a fleeting network interruption, or your remote write destination needs to be restarted. Prometheus will automatically retry sending data to the remote write destination at regular intervals. The initial interval is determined by the `min_backoff` setting (defaults to 30 ms), and it is doubled each subsequent time the request fails up to `max_backoff`, which defaults to 5 seconds. Once the `max_backoff` is reached, the request will continue to be retried every `max_backoff` interval until it succeeds.

Generally, the backoff settings should not need to be tweaked unless your remote write destination is sensitive to rate limits and you want to wait a longer period of time before your initial retry.

Now that we have an idea of how to tune remote write settings if you already have a remote write destination, let's start looking at popular remote write destination systems and set them up in our Kubernetes cluster.

Using VictoriaMetrics

VictoriaMetrics is one of the most popular and established options for the remote storage of Prometheus metrics. In fact, some people and projects use VictoriaMetrics as a drop-in replacement for Prometheus since it supports directly scraping targets via its `vmagent` component and is mostly compatible with the Prometheus configuration file (it lacks support for the `remote_write`, `remote_read`, `rule_files`, and `alerting` sections).

One of the primary reasons that people choose VictoriaMetrics as their remote storage destination is its resource efficiency. In a head-to-head benchmark against Prometheus, VictoriaMetrics boasts seven times the reduction in disk space (~0.3 bytes per sample vs. ~2.1 bytes) and five times the reduction in memory (~4.3GB of RAM vs. ~23GB) for the same amount of time series and samples (2.8 million active time series and 280,000 samples/sec) in their benchmark, published at `https://valyala.medium.com/prometheus-vs-victoriametrics-benchmark-on-node-exporter-metrics-4ca29c75590f`. Note, though, that this benchmark is from November 2020 and Prometheus has since had several improvements targeted at improving resource usage, such as the change to the `stringlabels` data structure, which reduces memory usage.

However, VictoriaMetrics is not without its drawbacks. In my mind, the two potential drawbacks to VictoriaMetrics are its approach to open source and its PromQL compatibility.

The core of VictoriaMetrics is open source and it is available on GitHub at `https://github.com/VictoriaMetrics/VictoriaMetrics/`. This provides the majority of the features that VictoriaMetrics is known for. Nevertheless, VictoriaMetrics is an independent company that needs to make money and, therefore, has a paid, closed source "enterprise" version of VictoriaMetrics that it sells to customers. This provides several features that are desirable to those running large, shared VictoriaMetrics environments, such as the downsampling of historical data, full-featured multi-tenancy, machine learning-based anomaly detection, and **mutual-TLS** (**mTLS**) between components when running in a clustered deployment.

In addition to withheld features, there is the ever-present concern of ongoing support for the open source product as a whole. VictoriaMetrics has been around as an open source project since May 2019 (when the single-node version was open sourced) and has been continuously updated since then. However, as we have seen in recent years with projects such as Elasticsearch and Terraform, both the license terms and development of an open source project supported by a for-profit company can change at any time. Consequently, there is an ever-present inherent risk when using open source software that depends upon the goodwill of a for-profit company to continue supporting it.

The other caveat to be aware of when considering VictoriaMetrics is that it is explicitly not 100% compatible with PromQL and does not aim to be. VictoriaMetrics implements its own query language, **MetricsQL**, which is intended to be a superset of PromQL. It implements all of the same query functions as PromQL and is fully compatible syntactically. VictoriaMetrics also implements the same REST API endpoints as Prometheus, so it can be used as a "Prometheus" data source in Grafana. However, the data returned by queries may differ when querying VictoriaMetrics vs. Prometheus directly. Based on benchmarks run on Prometheus v2.30.0 and VictoriaMetrics v1.67.0, VictoriaMetrics only achieved a 72.78% compliance score (see `https://promlabs.com/promql-compliance-tests/`).

> **Important note**
>
> For a full explanation of failed compliance tests, consult the Medium blog post linked at the end of this chapter.

Rather than being due to a lack of features, all of the "failed" tests were due to slightly different results due to intentional choices in MetricsQL. For example, when using some query functions, such as `round` or `predict_linear`, VictoriaMetrics will still include the metric name in the results, whereas Prometheus will remove it. Another common difference is the implementation of the `rate` and `increase` functions in MetricsQL. MetricsQL behavior differs from PromQL in that it also takes into account the last sample value from before the start of the specified interval, whereas PromQL only considers the values within the interval.

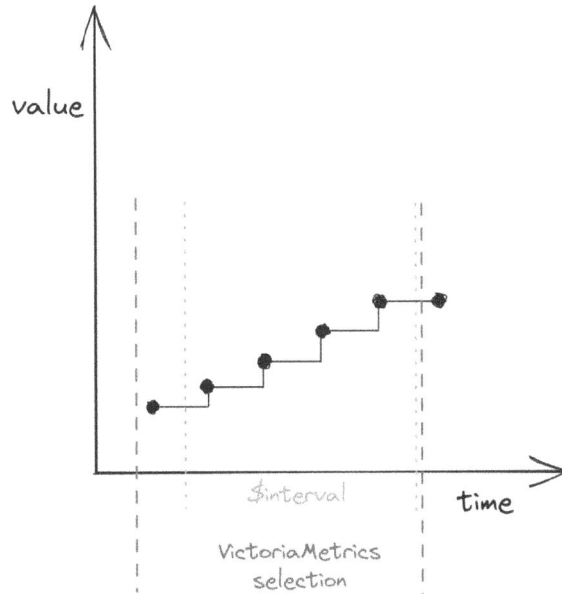

Figure 9.1 – VictoriaMetrics range selection behavior

As you can see from *Figure 9.1*, VictoriaMetrics will look back to the last data point prior to the range interval and include this when calculating the `rate` or `increase` of a time series, whereas Prometheus only includes data within the interval's range. Since the data being considered in these functions differs, they are likely to produce different results. Realistically, this should not matter in practice but may provoke questions from other users as to why their query results differ based on where they run them.

Now that we understand a little more about VictoriaMetrics in theory, let's experiment with it in a more hands-on approach by deploying it.

Deployment methods

VictoriaMetrics has two different methods of deployment: **single-node** and **clustered**. Both are available in the open source version of VictoriaMetrics and are targeted toward different use cases. If you're like me, you probably skipped right to "clustered" and already think that that is the way to go. After all, clustered means better, right? Before we go discounting the single-node version as inferior, let's take a look at what it is capable of.

According to VictoriaMetrics' docs (`https://docs.victoriametrics.com/Single-server-VictoriaMetrics.html#capacity-planning`), based on their case studies, they've found that a single-node VictoriaMetrics deployment is able to scale tremendously. They write the following:

A single-node VictoriaMetrics works perfectly with the following production workload according to our case studies:

- *Ingestion rate: 1.5+ million samples per second*

- *Active time series: 50+ million*

- *Total time series: 5+ billion*

- *Time series churn rate: 150+ million of new series per day*

- *Total number of samples: 10+ trillion*

- *Queries: 200+ qps*

- *Query latency (99th percentile): 1 second*

That is a tremendous scale for a single node to be able to handle. Consequently, unless the scale of your infrastructure is incredibly large, I think that the single-node version is more than sufficient for the majority of use cases. The primary reason to use clustered deployment is if you're building out a centralized, multi-tenant metrics database using it (similar to what `Fly.io` describes in `https://fly.io/blog/measuring-fly/`).

The clustered version of VictoriaMetrics adds additional operational complexity since it requires running multiple independent components, which also adds additional resource overhead. Consequently, even the VictoriaMetrics project itself strongly encourages users to vertically scale a single-node VictoriaMetrics instance as much as they can before considering the clustered deployment method.

With this in mind, let's deploy a single-node VictoriaMetrics instance using our Kubernetes cluster from *Chapter 2* and configure our Prometheus instance to remote write to it.

Deploying to Kubernetes

The VictoriaMetrics project has a variety of Helm charts available at `https://github.com/VictoriaMetrics/helm-charts/`, including one that will set up a VictoriaMetrics Kubernetes operator and another that is essentially the VictoriaMetrics-equivalent of the kube-prometheus stack we've already installed. However, we'll just be using the `victoria-metrics-single` chart, which will give us a single-node deployment without installing the Kubernetes **custom resource definitions** (**CRDs**) used by the operator.

To begin, we must first add the VictoriaMetrics Helm repository:

```
$ helm repo add vm https://victoriametrics.github.io/helm-charts/
$ helm repo update
```

Next, we'll need to configure the values that we want to pass to the Helm chart. The defaults are mostly sane for our use case, so we don't need to change much. We'll just disable the creation of a persistent volume for storing data since we are just testing this and will tear it down before the chapter's over:

```
server:
  persistentVolume:
    enabled: false

  scrape:
    enabled: false
```

Now we can apply the Helm chart, and we'll put it in the same namespace as our Prometheus instance:

```
$ helm install \
    --namespace prometheus \
    --values mastering-prometheus/ch9/vm-values.yaml \
    --version 0.9.10 \
    vmsingle \
    vm/victoria-metrics-single
```

After installing, the Helm chart should output some information on how to use VictoriaMetrics in your cluster, which looks like the following:

```
[. . .]
Metrics Ingestion:
  Get the Victoria Metrics service URL by running these commands in
the same shell:
    export POD_NAME=$(kubectl get pods --namespace prometheus -l
"app=server" -o jsonpath="{.items[0].metadata.name}")
    kubectl --namespace prometheus port-forward $POD_NAME 8428

  Write url inside the kubernetes cluster:
    http://vmsingle-victoria-metrics-single-server.prometheus.svc.
cluster.local:8428/api/v1/write
[. . .]
```

Make a note of that write URL since we'll need it when we configure Prometheus to send data to VictoriaMetrics. If you've been following along, your write URL should look the same as mine in the preceding example.

To confirm that VictoriaMetrics is up and running, we can navigate to its web UI by port-forwarding and visiting it in our local internet browser:

```
$ kubectl port-forward vmsingle-victoria-metrics-single-server-0 8428
```

Now, we can navigate to `http://localhost:8428/vmui` to the query UI for VictoriaMetrics. If you omit `/vmui` from the URL path, you can also see a landing page displaying the various other endpoints that VictoriaMetrics exposes. This isn't very useful without any data in VictoriaMetrics yet, so let's configure our Prometheus to start sending data.

In *Chapter 6*, we made our Prometheus deployment **highly available** (**HA**) by deploying multiple replicas of Prometheus scraping the same targets. As part of this, they're configured with an external label that identifies the replica:

```
global:
  external_labels:
    prometheus: prometheus/mastering-prometheus-kube
    prometheus_replica: prometheus-mastering-prometheus-kube-0
```

Unfortunately, this is incompatible with how VictoriaMetrics handles the deduplication of metrics from HA Prometheus instances. VictoriaMetrics requires that HA Prometheus instances be configured with **identical** external labels. Consequently, we'll need to remove `prometheus_replica` in addition to configuring Prometheus for remote write. We're able to accomplish this by setting `replicaExternalLabelNameClear` to `true` in our Helm values for the `kube-prometheus-stack` chart.

Additionally, we'll enable the provisioning of dashboards related to monitoring Prometheus remote write performance and define a remote write configuration to send all data to VictoriaMetrics.

To configure remote write in Prometheus, our Helm values should look like this now:

```
grafana:
  enabled: true
  defaultDashboardsTimezone: browser
  adminUser: root
  adminPassword: m@ster1ngPr0m3th3us
prometheus:
  prometheusSpec:
    serviceMonitorSelectorNilUsesHelmValues: false
    replicas: 2
    replicaExternalLabelNameClear: true
    remoteWriteDashboards: true
    remoteWrite:
      - name: victoriametrics
```

```
        url: http://vmsingle-victoria-metrics-single-server.
prometheus.svc.cluster.local:8428/api/v1/write
  cleanPrometheusOperatorObjectNames: true
```

Then, we can upgrade our Prometheus deployment:

```
$ helm upgrade --namespace prometheus \
    --version 47.0.0 \
    --values mastering-prometheus/ch9/prom-vm-values.yaml \
    mastering-prometheus \
    prometheus-community/kube-prometheus-stack
```

Once that's rolled out, you can go back to the VictoriaMetrics UI and begin querying for metrics to see the data coming in. Simple as that!

I'd encourage you to continue poking around in the VictoriaMetrics interface to experiment with some of the nice features it includes, such as its enhanced cardinality explorer and query tracing.

When you're ready to tear down your VictoriaMetrics instance, run the following Helm commands to clean it up and remove the remote write configuration from Prometheus (the referenced Helm values file is available in this book's GitHub repository):

```
$ helm uninstall vmsingle --namespace prometheus
```

```
$ helm upgrade --namespace prometheus \
    --version 47.0.0 \
    --values mastering-prometheus/ch9/prom-values.yaml \
    mastering-prometheus \
    prometheus-community/kube-prometheus-stack
```

With your VictoriaMetrics deployment cleaned up, we can move forward to seeing how Grafana Mimir solves the same problem in a slightly different way.

Using Grafana Mimir

Grafana Labs' **Mimir** project is a relative newcomer to the remote storage Prometheus ecosystem. Announced in 2022, Mimir is the spiritual successor to the popular Cortex project, which is a **Cloud Native Computing Foundation** (CNCF)-sponsored project (as with Prometheus, Thanos, Kubernetes, and many more). It aims to be the most scalable and performant option available for time series storage.

> Important note
>
> Is Cortex dead? Cortex is still under active development and is used extensively by numerous companies. If Mimir intrigues you, but its AGPLv3 license is off-putting, try out Cortex! It'll still be a mostly similar experience and architecture.

Note that Mimir does not mention being the most efficient in its goals. If you approach Mimir thinking that you are going to see the same massive gains in memory and storage efficiency as VictoriaMetrics, you will be disappointed. However, you still can expect its storage efficiency to be about as good as using Prometheus directly (roughly 2 bytes of space per sample). While efficiency gains will surely come in time as Mimir continues to mature as a project, it will certainly require more resources than running vanilla Prometheus due to the fact that it is doing more behind the scenes.

Comparing to VictoriaMetrics

Personally, when I think of VictoriaMetrics vs. Mimir, the main difference that matters to me is how it stores its data. Both can run as single-node or clustered deployments, both can scale read and write paths independently, and both can accept Prometheus data and query it with PromQL expressions. However, VictoriaMetrics only supports writing data to locally attached storage (local disks or block storage volumes such as AWS's EBS), whereas Mimir only supports writing data to object storage (such as S3, GCS, Azure Blob Storage, etc.).

> **Important note**
>
> Mimir technically supports writing data to a local disk if you're running in its single-node "monolithic" mode. However, this is considered an uncommon deployment type and is mostly recommended only for evaluation purposes.

Object storage is significantly cheaper than block storage at every major cloud provider. Consequently, you can store significantly more time series data for the same price in almost all cases. This enables much longer-term storage of metric data (think years instead of months), which can be a game changer depending on how your company is leveraging your Prometheus data. However, there is the caveat that VictoriaMetrics does store data much more efficiently, so some costs may be offset by that. Regardless, this seemingly minor implementation detail can have major implications for the cost and scalability of your remote storage architecture.

With regard to the performance differences between VictoriaMetrics and Mimir, VictoriaMetrics is—as expected—much more performant. Using a similar testing methodology to the previously discussed benchmark comparing VictoriaMetrics to vanilla Prometheus, VictoriaMetrics reports similar improvements over Mimir. VictoriaMetrics reports a >3x improvement in storage efficiency, a 1.7x CPU improvement, and a 5x memory improvement. However, Mimir did win out in median query latency, so it's not all bad news for Mimir.

We already knew that Mimir would be less efficient, though, right? Those tradeoffs are worth it to some people, so let's forge ahead with deploying our own Mimir to experiment with.

Deploying to Kubernetes

As mentioned previously, Mimir—like VictoriaMetrics—can be deployed either in its clustered "microservices" mode or a single-node "monolithic" mode. Unlike VictoriaMetrics, however, the monolithic mode isn't really recommended for production workloads. Nevertheless, we're just evaluating Mimir with our tiny Prometheus installation, so we'll give the monolithic mode a shot!

We'll once again configure both of our Prometheus replicas to send data to this remote storage backend. Unlike VictoriaMetrics, we do not have to make the external labels of our Prometheus instances unique. Instead, Mimir keeps track of highly available Prometheus replicas by electing a leader amongst them from which it will accept data. Data from the non-elected Prometheus replicas will simply be dropped. If the elected Prometheus stops sending data for 30 seconds (this is configurable), then Mimir will fail over to another replica as the new leader. Unfortunately, this will unavoidably result in some lost data.

In order to co-ordinate the deduplication of metric data, Mimir requires a **key-value** (**KV**) store of either Hashicorp's Consul or etcd to keep track of the elected leader across clustered components. For our experimental deployment, we can use an in-memory store that will keep track of Prometheus replicas, but this is not feasible for any production deployment using more than one replica of Mimir.

As if to reinforce my point that monolithic mode really is not a recommended way of running Mimir in production, Grafana Labs does not—at the time of writing—publish a Helm chart that supports deploying Mimir in this way. Consequently, we'll be using good old-fashioned hand-written Kubernetes `Deployment` and `Service` manifests.

> **Important note**
>
> Due to the length of these files, the contents you see below are truncated and not fully valid Kubernetes manifests. To see the full manifests for the Mimir `Deployment` and `Service`, consult the folder for this chapter in the GitHub repository for this book.

To deploy it, we'll use a file named `mimir.yaml` with contents that look roughly like the following:

```yaml
apiVersion: apps/v1
kind: Deployment
spec:
  replicas: 1
    spec:
      containers:
        - image: grafana/mimir:2.9.2
          name: mimir
          args:
            - "-common.storage.filesystem.dir=/data"
            - "-ruler.rule-path=/data/ruler"
            - "-blocks-storage.filesystem.dir=/data/ingester"
```

```
          - "-ingester.ring.replication-factor=1"
          - "-distributor.ha-tracker.enable-for-all-users=true"
          - "-distributor.ha-tracker.store=inmemory"
          - "-distributor.ha-tracker.enable=true"
          - "-distributor.ha-tracker.cluster=prometheus"
          - "-distributor.ha-tracker.replica=prometheus_replica"
```

It's definitely not as straightforward as the VictoriaMetrics deployment, but that is to be expected. Mimir has a litany of configurable options, which makes it extremely versatile for a variety of use cases and workload types. However, that also makes keeping track of all the command flags a bit tedious. For example, look no further than the five different flags we had to set to enable the de-duplication of metrics from HA Prometheus pairs.

Nevertheless, let's deploy this to our Kubernetes cluster:

```
$ kubectl apply -f mastering-prometheus/ch9/mimir.yaml
```

You can confirm that the Mimir pod is created and running with the following:

```
$ kubectl get pods --namespace prometheus
```

Finally, we once again need to configure Prometheus to send data to Mimir. One of the unique things about Mimir is that you will need to configure Prometheus to set an X-Scope-OrgID HTTP header on all requests it sends to Mimir to identify the Mimir "tenant" that the data belongs to. Similarly, the Grafana data source for Mimir also needs to specify that header when querying data. Oh, by the way, you need to set up a Grafana data source for Mimir to be able to query it from a web UI; it does not have a custom UI like VictoriaMetrics. Thankfully, our kube-prometheus-stack Helm chart can take care of provisioning that data source for us.

To configure Prometheus and Grafana for Mimir, our Helm values file should look like the following:

```
grafana:
  enabled: true
  defaultDashboardsTimezone: browser
  adminUser: root
  adminPassword: m@ster1ngPr0m3th3us
  additionalDataSources:
    - name: Mimir
      type: prometheus
      access: proxy
      orgId: 1
      url: http://mastering-prometheus-mimir.prometheus.svc.cluster.
local/prometheus
      version: 1
```

```
      editable: true
      jsonData:
        httpHeaderName1: "X-Scope-OrgID"
      secureJsonData:
        httpHeaderValue1: "mastering-prometheus"
prometheus:
  prometheusSpec:
    serviceMonitorSelectorNilUsesHelmValues: false
    replicas: 2
    remoteWriteDashboards: true
    remoteWrite:
      - name: mimir
        url: http://mastering-prometheus-mimir.prometheus.svc.cluster.
local:80/api/v1/push
        headers:
          X-Scope-OrgID: mastering-prometheus
cleanPrometheusOperatorObjectNames: true
```

We'll apply these new values with Helm:

```
$ helm upgrade --namespace prometheus \
    --version 47.0.0 \
    --values mastering-prometheus/ch9/prom-mimir-values.yaml \
    mastering-prometheus \
    prometheus-community/kube-prometheus-stack
```

This should immediately begin sending data to Mimir from both of our Prometheus replicas and add the new data source to our Grafana instance. To confirm, we can port-forward to Grafana and poke around by running queries in the **Explore** tab and by switching the data source on dashboards to Mimir to confirm that data is being ingested.

```
$ kubectl port-forward svc/mastering-prometheus-grafana 3000:80
```

Important note

It may take a minute or two for data to start showing up.

Once data is being sent, querying in Grafana looks the same as querying any other Prometheus datasource:

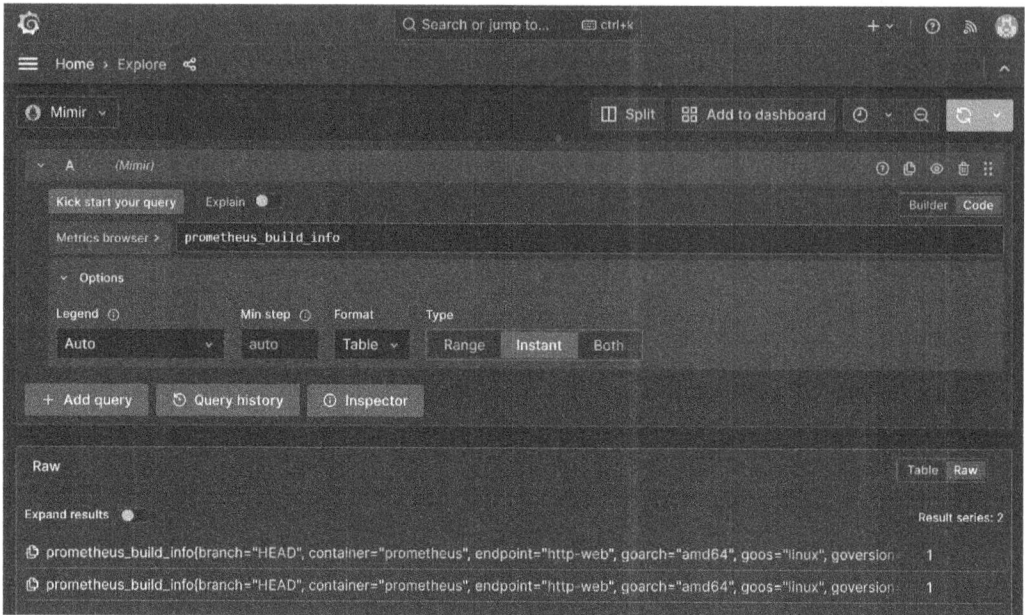

Figure 9.2 – Querying Mimir in Grafana

When you're ready to tear down your Mimir instance, run the following commands to clean it up and remove the remote write configuration from Prometheus (the referenced `helm values` file is available in this book's GitHub repository):

```
$ kubectl delete -f mastering-prometheus/ch9/mimir.yaml

$ helm upgrade --namespace prometheus \
    --version 47.0.0 \
    --values mastering-prometheus/ch9/prom-values.yaml \
    mastering-prometheus \
    prometheus-community/kube-prometheus-stack
```

If saying goodbye to your shiny new tools is hard, there's certainly more to explore with both Mimir and VictoriaMetrics. So, feel free to examine the different settings available. Tweak, tune, and experiment, then come back when you're ready to move on to the next chapter.

Summary

In this chapter, we learned about how Prometheus can integrate with a variety of remote storage systems through the remote read and remote write protocols. We made that practical by deploying two popular remote storage backends: VictoriaMetrics and Mimir. Both can be used to extend Prometheus far beyond what simple federation is capable of in order to support advanced use cases involving the multi-tenancy and long-term storage of Prometheus metrics.

In our next chapter, we're going to look at another popular open source project that integrates with Prometheus: Thanos. It integrates with remote storage protocols but also can be used for a whole lot more.

Further reading

To learn more about the topics that were covered in this chapter, take a look at the following resources:

- *Prometheus agent*: `https://prometheus.io/blog/2021/11/16/agent/`
- *Remote write configuration and tuning*:

  - `https://prometheus.io/docs/prometheus/latest/configuration/configuration/#remote_write`

  - `https://prometheus.io/docs/practices/remote_write/#remote-write-tuning`

- *Prometheus "stringlabels"*: `https://prometheus.io/blog/2023/03/21/stringlabel/`

- *VictoriaMetrics*:

  - `https://docs.victoriametrics.com`

  - `https://valyala.medium.com/prometheus-vs-victoriametrics-benchmark-on-node-exporter-metrics-4ca29c75590f`

  - `https://docs.victoriametrics.com/Articles.html#benchmarks`

  - `https://victoriametrics.com/plans-features/`

  - `https://medium.com/@romanhavronenko/victoriametrics-promql-compliance-d4318203f51e`

- *Mimir*:

  - https://grafana.com/blog/2022/03/30/announcing-grafana-mimir/

  - https://victoriametrics.com/blog/mimir-benchmark/

  - https://grafana.com/docs/helm-charts/mimir-distributed/latest/run-production-environment-with-helm/

  - https://grafana.com/docs/mimir/latest/configure/configure-high-availability-deduplication/

  - https://grafana.com/docs/mimir/latest/

  - https://o11y.tools/mimircalc/ (Mimir resource requirements calculator)

10

Extending Prometheus Globally with Thanos

The remote storage systems that we covered in the previous chapter are not for everyone. Perhaps you are perfectly happy to run your Prometheus instances globally without any centralized place where you're aggregating those metrics. However, it sure would be nice to have some way to run queries from a centralized place and fan them out to all of your Prometheus instances… Good news! Thanos can do that, and more!

Thanos is less of a pre-built, comprehensive solution the way VictoriaMetrics or Mimir are, and more of a Swiss Army knife of *a la carte* components that can be mixed and matched to fit your specific use case. At the time of writing, seven different components comprise the Thanos project and you can run as few or as many of them as you need.

In this chapter, we'll cover them all in these main topics:

- Overview of Thanos
- Thanos Sidecar
- Thanos Compactor
- Thanos Query
- Thanos Query Frontend
- Thanos Store
- Thanos Ruler
- Thanos Receiver
- Thanos tools

Let's get started!

Technical requirements

For this chapter, we'll continue building off of the Prometheus environment we deployed to Kubernetes in *Chapter 2*. Consequently, you'll need the two following tools installed:

- **kubectl**: `https://kubernetes.io/docs/tasks/tools/#kubectl`

- **helm**: `https://helm.sh/docs/intro/install/`

This chapter's code is available at `https://github.com/PacktPublishing/Mastering-Prometheus`.

Overview of Thanos

The Thanos project began at Improbable and was spearheaded by two scions of the Prometheus ecosystem, *Bartłomiej Płotka* and *Fabian Reinartz*, who have both also been significant contributors to Prometheus itself and various other Prometheus-related projects. After a short period, the project was donated to the **Cloud Native Computing Foundation** (**CNCF**), where it is now designated as an "incubating" project.

The three stated goals of the Thanos project are as follows:

- Global query view of metrics

- Unlimited retention of metrics

- High availability of components, including Prometheus

The *core* of Thanos (its original components) is comprised of Thanos Sidecar, Thanos Store, Thanos Compact, and Thanos Query. Each component is a subcommand of the `thanos` binary – no need to download and deploy separate executables for each of the components. Through each of its various components, Thanos enables distributed querying of HA Prometheus instances, long-term retention of metrics data in object storage, acting as a remote write receiver, and evaluation of recording and alerting rules across data sets from multiple Prometheus instances.

It's worth noting that, at the time of writing, Thanos has not released version 1.0 of the project. Consequently, there are no stability guarantees between minor versions. However, in my experience, breaking changes are few and far between in Thanos now. In the last eight minor version releases, breaking changes have mostly centered around renaming the metrics exposed by Thanos. As always, it is a good idea to check the release notes of new versions to see the specific changes and if they affect you.

> **Note**
> All discussion of Thanos for the rest of this chapter is based on v0.32.5 of Thanos, and specifics may have changed depending on when you are reading this.

Why use Thanos?

While Thanos may seem more complex at first glance since there seem to be so many components, it's the simplest of all of the projects we've looked at thus far. The multiple components are simple and easy to understand since their functionality is narrowly scoped. You also don't need to run all of the components. I can count on one hand the number of people I've met who are actively using *all* of the Thanos components in their environment.

With regards to support and longevity, Thanos is a well-established project with a healthy open source community of hundreds of contributors (I'm one of them!). Being a CNCF project, there's no risk that its license or feature set will be modified to encourage you to pay to use it. Support is also easy to come by through its Slack channel or other online forums.

So, why use Thanos? Because it can grow with you as you continue to expand your Prometheus environment. There's minimal effort to get started and it's easy to introduce new components as you go along. For that reason, I always recommend Thanos as the first place for people to start when they're looking to extend their Prometheus environment with additional features such as distributed querying or long-term data retention.

If you're convinced – as I hope you are – that you could benefit from Thanos, let's start exploring and deploying each component.

Thanos Sidecar

Thanos Sidecar is the most fundamental Thanos component, enabling Thanos's two most popular features: querying multiple Prometheus instances from a centralized location (Thanos Query) and backing up Prometheus TSDB data in an object storage backend.

Thanos Sidecar works by running alongside the Prometheus server as a "sidecar" (clever naming, huh?). This is technically only a strict requirement if you're using the sidecar to upload Prometheus data to object storage, but you should strive to deploy the Sidecar alongside your Prometheus instance, even if you don't use that feature. Doing so will help minimize latency between the Sidecar and Prometheus.

The Sidecar fulfills its first job of enabling distributed querying of Prometheus instances by exposing a gRPC API that Thanos Query (or other Thanos components such as Ruler) can communicate with. This gRPC API (henceforth referred to as `StoreAPI`) is implemented and exposed by the Sidecar, Ruler, Receiver, Store, and Query components. It consists of only four different RPCs: `Info`, `Series`, `LabelNames`, and `LabelValues`.

Thanos Query communicates with the Sidecar over the StoreAPI to proxy requests to the Prometheus instance that the Sidecar runs alongside. The Sidecar receives these requests and then uses Prometheus' Remote Read API (see *Chapter 9*) to execute a PromQL query and return the results. Since this functionality essentially just uses Thanos Sidecar as a proxy, it can work without being co-located with the Prometheus instance – you could run Thanos Sidecar on your laptop and point its `--prometheus.url` flag to any Prometheus instance you like.

The other main functionality of the Sidecar is to send Prometheus data to an object storage service for long-term storage. For this to work, your Thanos Sidecar *must* be running alongside your Prometheus instance because it requires access to the data directory where Prometheus stores its TSDB blocks. Rather than solutions such as Mimir or VictoriaMetrics, where you send your Prometheus data via Remote Write to be stored according to that software's specifications, Thanos will simply take the TSDB blocks that Prometheus has written and upload them to object storage for safekeeping:

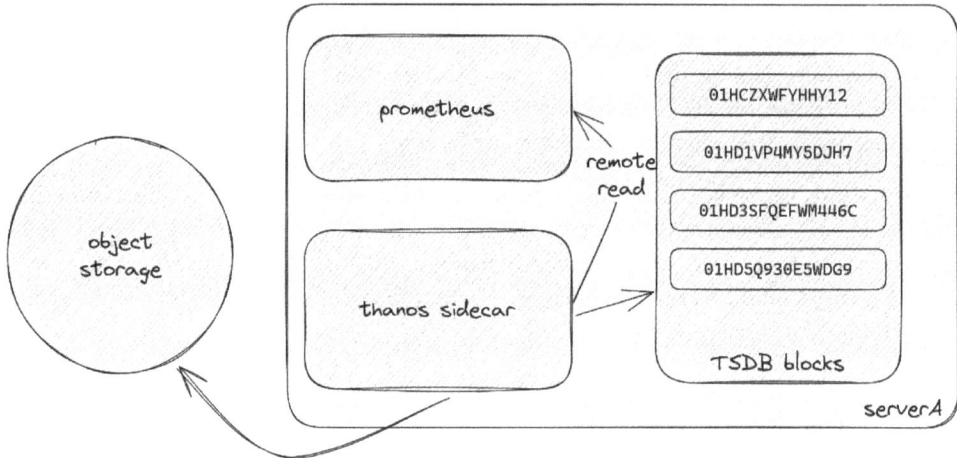

Figure 10.1 – Thanos Sidecar

For this to work properly, though, you must disable Prometheus's built-in compaction to prevent Prometheus from compacting multiple small blocks into a large one. This can be accomplished by setting Prometheus's `--storage.tsdb.min-block-duration` and `--storage.tsdb.max-block-duration` flags to equal values (2h is recommended). Thanos will ignore any Prometheus block with a `compaction.level` field value of more than 1 in its `meta.json` file.

> **Uploading compacted blocks**
>
> Even though Thanos will not upload compacted blocks by default, you can disable this behavior by setting the `--shipper.upload-compacted` flag on the Sidecar. This is intended to be used solely for uploading historical data as a one-time operation. Once the existing compacted blocks are uploaded, the flag should be removed from the Sidecar.

The final functionality that the Sidecar provides is the ability to automatically reload your Prometheus instance when certain files change on disk. So long as you pass the `--web.enable-lifecycle` flag to the Prometheus instance, Thanos can tell Prometheus to reload its configuration when Thanos Sidecar detects that the Prometheus configuration file (specified by Thanos's `--reloader.config-file` flag) changed or one or more Prometheus rules files (specified by Thanos's repeatable `--reloader.rule-dir` flag) changed.

When using the reloader feature of Thanos Sidecar, you can also have Thanos perform variable substitution on the configuration file before reloading Prometheus. Variables in `$(VARIABLE_NAME)` format will be substituted based on environment variables from the server/container Thanos Sidecar is running in, which can be useful for substituting secret values into the configuration.

By specifying Thanos Sidecar's `--reloader.config-envsubst-file` flag, Thanos will read the configuration file specified by `--reloader.config-file` and then write it out to the path specified by `--reloader.config-envsubst-file`. In this setup, Prometheus should be configured to use the file specified by `--reloader.config-envsubst-file` as its config file.

Armed with the knowledge of how Thanos Sidecar works, let's get it deployed. Since its query feature relies on Thanos Query to use and its long-term storage feature relies on Thanos Store to retrieve data from object storage, we won't be able to fully see the value of the Sidecar until we get to those components later in this chapter. However, this will lay the groundwork that we'll leverage throughout the rest of this chapter, so let's get started.

Deploying Thanos Sidecar

We'll continue building off of the Prometheus stack we deployed with the `kube-prometheus-stack` Helm chart in *Chapter 2*. The Prometheus Operator has built-in support for running Thanos Sidecar, so this will be relatively straightforward.

Before we deploy the Sidecar, we'll need to create an object storage destination for Thanos Sidecar to upload TSDB blocks to. We'll continue using Linode services, but you can use any object storage, such as AWS S3, Azure Blob, or Google Cloud Storage (GCS).

We can use `linode-cli`, which we installed in *Chapter 2*, to create the Object Storage bucket, but first, we'll need to install the `boto3` dependency to interact with Object Storage:

```
$ pip3 install --user boto3
```

With that installed, we can create our bucket. I'm going to use the `us-ord-1` cluster, but feel free to pick whatever cluster you want from the output of running `linode-cli object-storage clusters-list`. Your bucket name *must* be unique across all customers within an OBJ region, so make sure that you pick a different bucket name and/or cluster than what's shown in this example:

```
$ linode-cli obj --cluster us-ord-1 \
    mb \
    mastering-prometheus-thanos
```

Once the bucket has been created, we'll need to create an access key to provide to Thanos Sidecar for it to interact with the bucket:

```
$ linode-cli object-storage \
    keys-create \
```

```
--label=mastering-prometheus-thanos \
--bucket_access.cluster=us-ord-1 \
--bucket_access.bucket_name=mastering-prometheus-thanos \
--bucket_access.permissions=read_write
```

Make note of the `access_key` and `secret_key` values that are output by the preceding command – we'll need to pass them to the Sidecar for it to interact with the bucket. We can accomplish this by creating a Kubernetes `Secret`.

Create a new file called `thanos-objstore-config.yaml` that looks like this (if you're following along in this book's GitHub repository, just edit the existing file in the directory for this chapter):

```yaml
apiVersion: v1
kind: Secret
metadata:
  name: thanos-objstore-config
  namespace: prometheus
stringData:
  thanos.yaml: |
    type: s3
    config:
      bucket: mastering-prometheus-thanos
      endpoint: us-ord-1.linodeobjects.com
      access_key: REDACTED
      secret_key: REDACTED
```

Substitute your `access_key` and `secret_key` values. If you used a different Object Storage cluster than `us-ord-1`, replace that value too (but keep the `.linodeobjects.com` suffix). Finally, apply it with `kubectl`, like so:

```
$ kubectl apply -f mastering-prometheus/ch10/thanos-objstore-config.
yaml
```

With that in place, we can update our Helm values to deploy Thanos Sidecar alongside our Prometheus instances. We only need to add a new few lines to get up and running:

```yaml
prometheus:
  prometheusSpec:
    thanos:
      objectStorageConfig:
        name: thanos-objstore-config
        key: "thanos.yaml"
  thanosServiceMonitor:
    enabled: true
```

> **Note**
>
> The preceding code only shows the new values we had to add. For the full Helm values used to deploy the Helm chart, consult the `ch10/sidecar-values.yaml` file in this book's GitHub repository, which was linked at the beginning of this chapter.

Note that we did not have to explicitly disable Prometheus's local compaction by setting the `--storage.tsdb.min-block-duration` and `--storage.tsdb.max-block-duration` flags to 2h each. This is because the Prometheus Operator will automatically do this when you enable Thanos Sidecar.

The `thanosServiceMonitor` section is also not strictly a requirement, but it creates a `ServiceMonitor` object for Thanos Sidecar so that Prometheus will discover and collect metrics from the Sidecar instances. The more metrics, the merrier.

Finally, apply the Helm chart with the updated values:

```
$ helm upgrade --namespace prometheus \
    --version 47.0.0 \
    --values mastering-prometheus/ch10/sidecar-values.yaml \
    mastering-prometheus \
    prometheus-community/kube-prometheus-stack
```

Now, we have a `thanos-sidecar` container running in all of our Prometheus pods! It may take a few hours before we see anything in the Object Storage bucket since Prometheus needs to write out a new, uncompacted block for the Sidecar to upload, but we're well on our way to greatness.

The Sidecar on its own isn't that useful, though. We need to get other Thanos components running to leverage it. First, let's take care of a brief housekeeping task – enabling compaction of uploaded TSDB blocks by setting up Thanos Compactor.

Thanos Compactor

The Thanos Compactor component is responsible for compacting and downsampling TSDB blocks stored in our Object Storage provider. Since we've disabled local compaction of TSDB blocks on the Prometheus instance, we still need to compact them somehow to ensure efficient storage of our data. Hence, the Thanos project provides a component for compaction.

Thanos Compactor handles compaction in the same way that Prometheus does – it takes several small blocks and compacts their indices and samples to make a larger block with an index that uses less space than if all the composite blocks still maintained a separate index. This relies on the presupposition that most time series exist across multiple sequential blocks, which should almost always be the case.

There's not much to note about how Thanos achieves this other than the requisite changes to account for the fact that the Compactor must download the blocks from object storage to compact them. However, since our Prometheus replicas are uploading to the same object storage bucket, you might be wondering how Thanos handles that situation. The answer lies in an optional feature known as **vertical compaction**.

Vertical compaction

Horizontal compaction is the type of compaction that we've exclusively seen thus far. Blocks do not overlap on the X-axis of time and are compacted based on the fact that they are sequential:

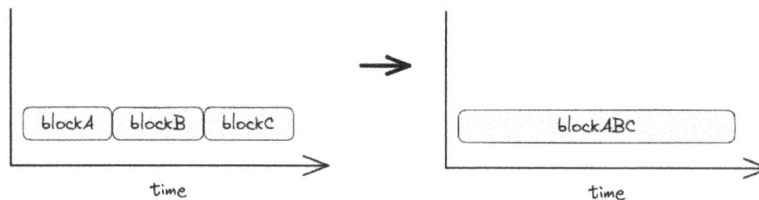

Figure 10.2 – Horizontal compaction

In vertical compaction, blocks overlap and are still compacted together based on some criteria. For example, let's say we have multiple Prometheus instances collecting data from the same scrape targets at the same intervals. Thanos Query (when we set that up) handles deduplicating results from HA Prometheus replicas at query time, but it may seem wasteful to be storing all that duplicated data in your object storage provider and at least double the storage you're using:

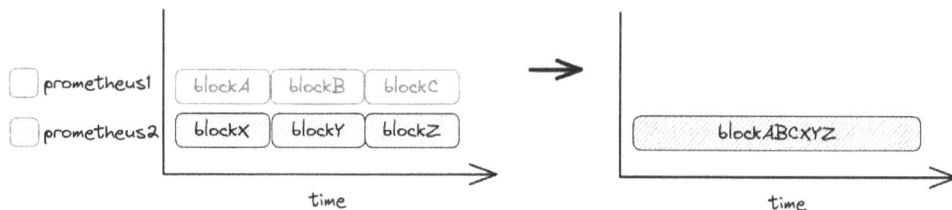

Figure 10.3 – Vertical Compaction

If you pass the `--compact.enable-vertical-compaction`, `--deduplication.replica-label=prometheus_replica`, and `--deduplication.func=penalty` flags to Thanos Compactor, then you can enable vertical compaction in the environment we've built out thus far.

Why so many flags? Well, vertical compaction is dangerous and potentially destructive. If you misconfigure it, you could seriously damage your historical data or even render it useless. As such, it is only recommended for advanced use cases where cost savings are paramount, and the risk is acceptable. Certainly do not enable it after *only* reading this book – be sure to review the official Thanos Compactor documentation to ensure you have a complete understanding of all of the associated risks: `https://thanos.io/v0.32/components/compact.md/#vertical-compaction-risks`.

Downsampling

The other core feature of the Compactor is downsampling. Downsampling is often misunderstood as another way in which to reduce the amount of storage required for historical metrics. On the contrary, using Thanos's downsampling feature will *increase* the amount of storage space that you use.

The primary function of Thanos's downsampling is to improve query performance when executing queries over long time ranges. This is accomplished by reducing the amount of data that needs to be retrieved from object storage through the consolidation of samples to a lower resolution. Thanos uses two non-configurable downsampling resolutions of 5 minutes (5m) and 1 hour (1h). This means that raw data is *downsampled* into one data point every 5 minutes and one data every hour, respectively.

The logic that Thanos uses to determine when to apply downsampling is as follows:

1. Raw data older than 40 hours is downsampled to the 5m resolution.

2. 5m resolution data older than 10 days gets downsampled to the 1h resolution.

When applying downsampling, new, additional TSDB blocks are created with the new resolutions. Consequently, in addition to your *raw* TSDB block directly from Prometheus, you'll also have a block with a 5m resolution and a block with a 1h resolution.

Per-resolution retention policies

Thanos allows you to configure different retention policies for different resolutions of data using the `--retention.resolution-raw`, `--retention.resolution-5m`, and `--retention.resolution-1h` flags. Unless you fully understand what you're doing, you should set these all to the **same value**. If you have different retention policies for different resolutions, you're likely to run into unpredictable query behavior.

For example, if you retain raw data and 5m resolution data for 90 days but 1h resolution data for 365 days, then your queries for data older than 90 days must have a query `step` of >=1h; otherwise, your query will return no results. This severely limits your ability to "zoom in" on historical data.

Even with the reduction in samples within the block, the index size remains roughly the same, so you can expect your storage usage to be at least double (probably triple) when utilizing downsampling. Take this into consideration when you're deciding if the query speedups and associated reduction in requests to object storage when querying large time ranges are worth the trade-off. If you don't want any downsampling at all (not recommended), it can be disabled via the `--downsampling.disable` flag on the Compactor.

Deploying Thanos Compactor

With the deeper respect and appreciation we now have of how the simple, modest Thanos Compactor has such a significant impact on our Thanos deployment, we can move forward with deploying it. To deploy it (and the remainder of our Thanos components), we'll be using manifests specific to this book based on the **kube-thanos** project (`https://github.com/thanos-io/kube-thanos`).

These manifests are not recommended for production usage, but perhaps you can use the Jsonnet files in this book's GitHub repository as a starting point to build manifests for production deployment in the future.

To deploy Thanos Compactor, all we need to run is the following `kubectl` command:

```
$ kubectl apply -f https://raw.githubusercontent.com/PacktPublishing/
Mastering-Prometheus/main/ch10/manifests/thanos-compact.yaml
```

After this, you should see a new `Pod` named `thanos-compact-0` running. Depending on how many blocks your Sidecar managed to upload thus far, it may take a while for the Compactor to complete its initial runs of compaction and downsampling. After completion, it will continue to run automatically every 5 minutes (configurable via the `--wait-interval` flag).

We'll set up Thanos Store later to pull data back out of object storage during querying. In the meantime, let's take a look at Thanos Query so that we can start querying our metrics via Thanos Sidecar.

Thanos Query

Thanos Query is another of the most fundamental Thanos components. Without it, there's not much point to Thanos Sidecar. It provides both a web UI and an API that are used to execute PromQL queries across multiple data sources (for example, Prometheus via Thanos Sidecar, metrics in object storage via Thanos Store, and so on).

The web UI will feel familiar to anyone who has used the Prometheus web UI since their functionality and UX are essentially equivalent. The query API is also 100% PromQL compliant and therefore can be used as a Prometheus-typed data source in Grafana. In practice at companies I've been at, we've even tended to use Thanos Query as our default data source in Grafana.

Thanos Query works by connecting to one or more **endpoints** that implement Thanos's gRPC-based StoreAPI. Endpoints can be specified via the repeatable `--endpoint` flag. This flag supports both static definitions (for example, `--endpoint=192.168.1.2:10901`) and dynamic discovery over DNS using the `dns+` and `dnssrv+` prefixes (for example, `--endpoint=dns+thanos-sidecars.mycompany.com:10901`).

The dns+ prefix will cause Thanos to look up the A/AAAA records for the provided domain name and add all returned IP addresses as endpoints. The dnssrv+ prefix, on the other hand, will look at SRV records and is especially useful in Kubernetes environments like ours where we can leverage the cluster DNS for Service via --endpoint=dnssrv+_grpc._tcp.prometheus-operated. prometheus.svc.cluster.local.

> **Fun fact**
>
> The Thanos Query component itself implements the StoreAPI and consequently, you can connect Thanos Query to other Thanos Query instances!

Deploying Thanos Query

To deploy Thanos Query, we will once again be leveraging manifests by leveraging the kube-thanos project that was built for this book.

All we need to run is the following kubectl command:

```
$ kubectl apply -f https://raw.githubusercontent.com/PacktPublishing/
Mastering-Prometheus/main/ch10/manifests/thanos-query.yaml
```

Now, we should have Thanos Query running and connected to our two Sidecars. It will also attempt to discover some of our yet-to-be-deployed Thanos components, so you can safely ignore any errors you see in the logs about that. You can experiment with the UI by port-forwarding to the service like so:

```
$ kubectl port-forward svc/thanos-query 9090
```

At this point, the web UI will be accessible in your browser at http://localhost:9090.

Scaling Thanos Query

Before we move on to the next component, let's briefly consider some scaling issues you may run into with Thanos Query. With a sufficient number of endpoints connected to Thanos Query, you will inevitably begin to see slowdowns in your queries as Thanos Query needs to fan out queries to more and more data sources, aggregate them locally, and display them. However, there are some steps you can take to minimize this and keep your Thanos Query instance(s) running in tip-top shape.

Custom PromQL engine

Historically – and still by default, at the time of writing – Thanos Query has used the same PromQL engine that Prometheus uses to evaluate its queries. However, this has several drawbacks to it, especially when running at a sufficiently large scale.

For one, the Prometheus PromQL engine is single-threaded, which limits the ability to vertically scale a Thanos Query instance to improve query speeds by adding more CPU cores. Most glaringly, though, is the fact that Thanos Query must pull *all* relevant series into memory locally before performing any aggregations on it. In other words, if you run a query such as sum(my_metric) on Thanos Query, it will poll all of its connected data sources to return my_metric before applying the sum operator on it, as opposed to running sum(my_metric) on each data source and then summing the results of that. While it may seem superfluous for aggregations such as sum, it *is* a necessary design to have statistically accurate results for other query aggregations such as avg. You couldn't push down the evaluation of the avg aggregation to the individual Prometheus instances since the average of multiple averages is not a true average.

There are some steps you can take to try to improve specific queries through features such as query sharding via the Query Frontend component or query pushdown. However, both of these are limited in their application and have diminishing returns. Consequently, I don't recommend spending too much time trying to get them working in your environment. Instead, the Thanos project has an exciting new development in which they have begun building their own, custom PromQL engine.

This PromQL engine is bundled in all recent versions of Thanos and can be optionally set as the default engine to use via the --query.promql-engine flag (which defaults to prometheus, though it can be set to thanos instead). It is multi-threaded and enables distributed execution of queries across Thanos Query instances. Recall earlier how we mentioned that you can connect Thanos Query to another Thanos Query as an endpoint? That's where this engine can shine:

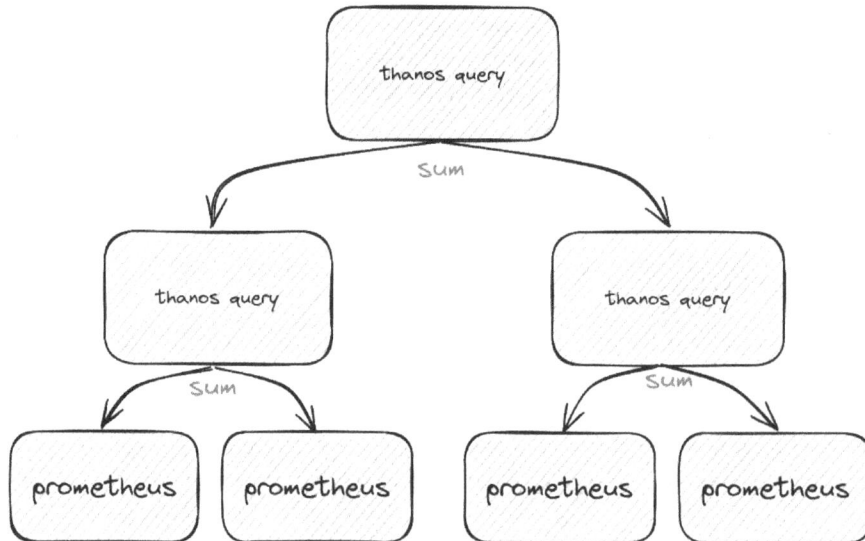

Figure 10.4 – Distributed query execution with Thanos's PromQL engine

With this distributed execution, a Query node can delegate portions of a query's execution to sub-Query nodes rather than having to pull all of the time series into its own memory. With our prior example of sum(my_metric), this engine would execute sum(my_metric) on the two sub-Query nodes, each of which would, in turn, pull the `my_metric` data from its two sub-Prometheus nodes and perform the `sum` operation. Then, the top-level querier would only need to `sum` the results from those two sub-Query nodes.

As of the end of 2023, the Thanos PromQL engine is still considered experimental. However, it should always fall back to using the default Prometheus PromQL engine if it encounters any issues executing a query. Try it out and see how it works in your environment!

Query Frontend

The other piece that is helpful when scaling Thanos Query is the addition of the Query Frontend component. This is an additional Thanos component that is put in front of Thanos Query to enable some neat features such as query parallelization and results caching. Since it is another individual Thanos component, we'll dive into it in more depth in the next section and deploy it.

Thanos Query Frontend

Thanos Query Frontend is a service that can be deployed in front of Thanos Query to improve query performance by splitting *large-range* queries into smaller ones and also caching results. It is based on a similar component implemented by Cortex (`https://github.com/cortexproject/cortex`), the predecessor to Mimir. You can think of it as a pre-processor of queries, where the majority of actual work is still done by the downstream queries.

Query sharding and splitting

Presuming you run multiple top-level Thanos Query instances, you can put Query Frontend in front of them to share the load between them more efficiently than simply load balancing between the two of them with something such as Nginx. This can be accomplished through **query splitting** based on time ranges and/or **vertical sharding**.

Query splitting

By default, the `--query-range.split-interval` flag is set to split range queries on a 24h interval. This means that if you query `sum(my_metric)` over the past week, Query Frontend would split that query into seven individual queries (each selecting 1d of data) and distribute those queries among the Thanos Query instances it is connected to:

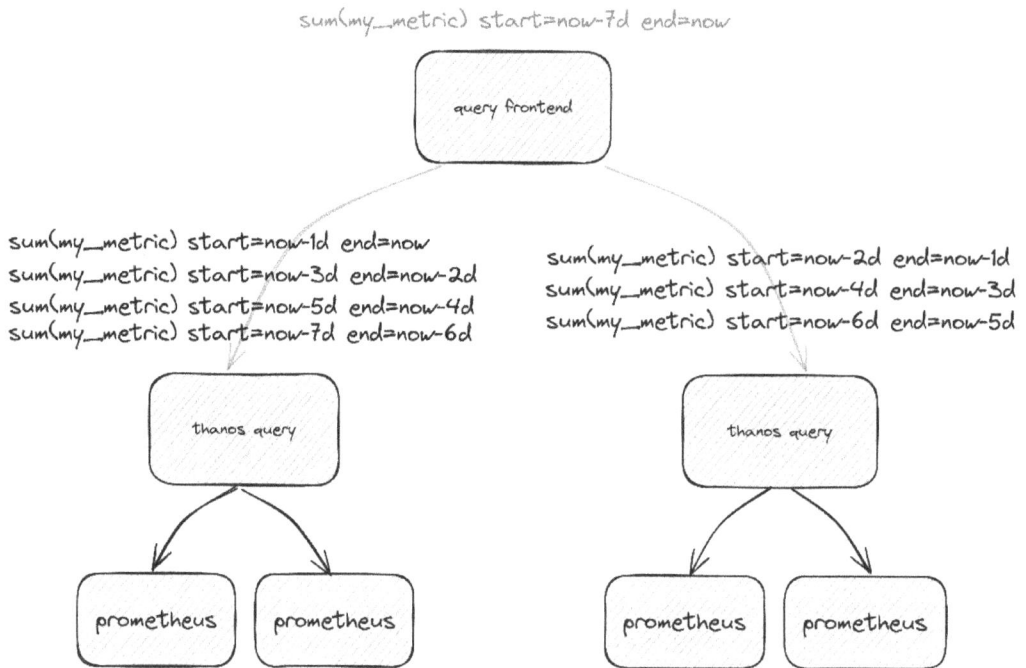

Figure 10.5 – Query Frontend time-based splitting

This time-based splitting enables better parallelization of queries and reduced overhead on downstream queries by splitting up large queries into more manageable chunks, which, in turn, reduces the risk of large queries causing individual Thanos Query instances to run **out of memory** (**OOM**). This can be a significant improvement for large environments with tens or hundreds of millions of time series.

Vertical sharding

In addition to splitting queries by time range, Thanos Query Frontend supports vertical sharding of specific queries. Vertical sharding entails breaking a query up into smaller pieces based on labels rather than by time ranges. Consequently, for vertical sharding to work, the query must include some aggregation operator such as by. This enables the frontend to propagate information to StoreAPI endpoints so that it only returns a subset of series for the relevant query.

For example, let's say we have the following series:

```
up{pod="prometheus-0", region="us-east", role="infra"}
up{pod="prometheus-0", region="jp-osa", role="infra"}
up{pod="prometheus-1", region="us-east", role="apps"}
up{pod="prometheus-1", region="jp-osa", role="apps"}
```

Using a vertical sharding factor of 2 (set by `--query-frontend.vertical-shards`) and a query of `count by (pod) (up)`, the query would be divided into two shards based on the value of the `pod` label.

This can help speed up queries and reduce memory usage of the data path but is limited in its application since it does require that some level of aggregation be performed. Consequently, you are still likely to see greater performance gains by architecting for and using the custom PromQL engine mentioned in the *Thanos Query* section since that can do more consistent sharding and delegation of queries.

Caching

In addition to dividing up queries, significant user experience gains can be gained from the addition of Query Frontend's result caching. For environments with frequently used Grafana dashboards, it can be wasteful to re-run the same queries over and over when multiple people are simultaneously viewing the dashboard. Query Frontend can cache results from queries and reuse them when subsequent queries are issued for the same data over an overlapping period. If a query requests data that is only partially cached (for example, if a query's result over the past 24 hours is cached but the query requests 48 hours), then the frontend can intelligently request only the non-cached data.

> **Note**
> Caching is only supported for range queries. Instant queries are not supported at the time of writing.

Caching can be accomplished through three different supported backends: in-memory, Memcached, and Redis. The in-memory cache only keeps the cache locally on Query Frontend, so it cannot be shared between multiple Query Frontend instances. Consequently, you're likely better off using Memcached or Redis in production.

In addition to caching query results, Query Frontend can also cache the results of the labels returned from StoreAPIs. This is used by features such as auto-complete in query builders in the Thanos and Grafana UIs to populate possible label names and values. However, it is configured separately from request caching (`--labels.response-cache-config-file` versus `--query-range.response-cache-config-file`), although both can use the same caching backend.

Deploying Thanos Query Frontend

Thanos Query Frontend is meant to be tuned to your specific use case and the types of queries that are most often run in your infrastructure. Consequently, it's not recommended that you run the example we'll use in this book in production. Additionally, we won't implement external caching through Memcached or Redis since that is outside the scope of this book. Instead, we'll just use the in-memory cache.

To deploy Thanos Query Frontend, we will once again be leveraging manifests leveraging the kube-thanos project that was built for this book.

All we need to run is the following `kubectl` command:

```
$ kubectl apply -f https://raw.githubusercontent.com/PacktPublishing/
Mastering-Prometheus/main/ch10/manifests/thanos-queryfrontend.yaml
```

You can then port-forward to Query Frontend (be sure to stop any prior port-forwards using the same port), like so:

```
$ kubectl port-forward svc/thanos-query-frontend 9090
```

Then, by going to `http://localhost:9090` in your browser, you'll be able to leverage Query Frontend in your queries. Experiment with a variety of queries over a variety of time ranges and check out the `thanos_frontend_split_queries_total` metric to see how often the frontend is splitting your queries and if they're returning faster as a result.

Performance gains may be harder to see in our environment due to its relatively small amount of metrics and cardinality. However, I was able to see results that were more than 33% faster (~20s versus ~30s) when running queries like this:

```
histogram_quantile(0.99, rate(etcd_request_duration_seconds_
bucket[5m]))
```

This was done over a 1-week time range versus running it directly on our Thanos Query instance.

Additionally, when scaling Thanos Query up, you may observe additional performance gains since we only deployed one replica in the previous section. When adding an additional Thanos Query replica into the mix, I was able to see queries return in half the time (~15s versus ~30s). You can alter the number of replicas by running the following:

```
$ kubectl scale deployment/thanos-query --replicas=2
```

Since the usage of Query Frontend can significantly impact the performance of the read path for your Prometheus metrics, I encourage you to experiment further with tweaks to see how you can extract the most value in your environment. However, it's now time to check back in on those metrics we've been shipping to object storage. Thanos Sidecar has been diligently shipping them and Thanos Compact has been diligently compacting and downsampling them, but how do we get that data back out of object storage? Through the use of our next component, Thanos Store!

Thanos Store

Thanos Store is perhaps the simplest Thanos component in terms of usage. There is not too much to tweak or tune and it does not require much in terms of resources. It is effectively stateless, although you can use persistent storage with it to reduce startup time as it populates metadata about the available blocks in object storage. We'll get into some ways that you can horizontally scale it later, but for now, let's just focus on what it is and how to deploy it.

Thanos Store is another component that implements the StoreAPI, so you can use Thanos Query to pull data from it. Thanos Store's purpose is to function as a gateway to object storage, which is why you may sometimes see it referred to as the "Store Gateway."

When Thanos Sidecar uploads blocks and when Thanos Compactor operates on them, they update the `meta.json` file within that block (see *Chapter 3* for more information on that file) with a new `thanos` section. Thanos Store can use that additional information to know what blocks it should pull from object storage to service a given query through a combination of checking external labels and checking the start/end times of the block:

```
"thanos": {
    "labels": {
        "prometheus": "prometheus/mastering-prometheus-kube",
        "prometheus_replica": "prometheus-mastering-prometheus-kube-1"
    },
    "downsample": {
        "resolution": 0
    },
    "source": "sidecar",
    "segment_files": [
        "000001"
    ],
    "files": [
        {
            "rel_path": "chunks/000001",
            "size_bytes": 11922877
        },
        {
            "rel_path": "index",
            "size_bytes": 6919579
        },
        {
            "rel_path": "meta.json"
        }
    ]
}
```

Figure 10.6 – The "thanos" section of the meta.json file

Essentially, Thanos Store will function to retrieve historical data on-demand from object storage when a query necessitates it. Notably, the data will still be retrieved from object storage, even if it's available in a local Prometheus instance via Thanos Sidecar. This is because Thanos Query will fan out queries to all connected StoreAPI endpoints that might have data relevant to the query, regardless of its component type.

If you're trying to minimize your requests to your object storage provider, I would recommend setting the `--max-time` flag on your Thanos Store to a value that minimizes overlap. For example, if your Prometheus instance is configured to store 7d of data locally, you could configure your Thanos Store with `--max-time="-7d"` to prevent it from attempting to pull data from the last 7 days out of object storage. Of course, you run the risk of missing data if your Prometheus instance is down or lost data for some reason, but you can always remove that flag if needed. Regardless, without further ado, let's get this thing set up!

Deploying Thanos Store

To deploy Thanos Store, we will once again be leveraging manifests while leveraging the kube-thanos project built for this book.

All we need to run is the following `kubectl` command:

```
$ kubectl apply -f https://raw.githubusercontent.com/PacktPublishing/
Mastering-Prometheus/main/ch10/manifests/thanos-store.yaml
```

Now, you can run queries against data from object storage, including downsampled data!

Scaling Thanos Store

Thanos Store, while certainly simple up-front, can become increasingly difficult to scale based on how much data you are storing in object storage and how frequently you are retrieving it. With years worth of data and frequent queries spanning weeks, months, and even years, it can quickly become a bottleneck to just have a single Thanos Store instance handling all of that work. Thankfully, Thanos Store can accomplish some level of partitioning and caching to assist with these scaling challenges.

Partitioning

Partitioning in Thanos Store is more manual than in Thanos Query Frontend. It's not something that – at this point – you can do automatically. Instead, you need to manually partition your Thanos Store instances through one of two methods (or both!): time-based or external label-based.

Time-based

Time-based partitioning in Thanos Store is the concept of running multiple Thanos Store instances where each is responsible for only a specific slice of time. For example, if retaining 90 days of data in object storage, you might run three Thanos Store instances that are responsible for 0-29 days ago, 30-59 days ago, and 60-90 days ago, respectively. This helps limit the prospective dataset that each instance is responsible for, thereby reducing the potential resource overhead coming from an individual instance. It also helps further parallelize the query by allowing multiple store servers to operate on independent datasets and return their partial results to the querier to assemble the result vector.

The time ranges that a Thanos Store instance will consider TSDB blocks for can be configured via the `--min-time` and `--max-time` flags on Thanos Store and can either be explicit (for example, `2021-10-06T20:06:32Z`) or relative (for example, `-30d`).

External label-based

In addition to (or in place of) time-based partitioning, you can also shard Thanos Store instances based on the external labels of TSDB blocks. Each TSDB block includes the external labels of the Prometheus instance from whence it originated in the `thanos` section, as we saw earlier in *Figure 10.6*. We can configure Thanos Store to only return data for blocks with specific external labels through the use of a configuration similar to the relabel configurations we've seen elsewhere in Prometheus.

Relabel configurations can be specified via the `--selector.relabel-config-file` or `--selector.relabel-config` flag and only support a limited set of operations: `keep`, `drop`, and `hashmod`. So, if we wanted to configure Thanos Store to only return blocks for a specific Prometheus replica, we could provide a relabel config like this:

```
- action: keep
  regex: "prometheus-mastering-prometheus-kube-0"
  source_labels:
    - prometheus_replica
```

Additionally, you can use the special `__block_id` label to do sharding based on the hashmod of that label, similar to how we configured Prometheus shards back in *Chapter 6*. That might look like this on Thanos Store:

```
- action: hashmod
  source_labels: ["__block_id"]
  target_label: shard
  modulus: 3
- action: keep
  source_labels: ["shard"]
  regex: 0
```

In the preceding example, we use a `modulus` value of 3 to divide blocks across three instances of Thanos Store; each Thanos Store will use a different regex (`0`, `1`, or `2`) to decide which blocks to keep.

Caching

With so many requests to object storage, it makes sense to cache data to reduce your requests, which is beneficial both from a cost perspective and a latency perspective. Thanos Store supports caching two different types of data: *index data* and *chunk data*. Similar to Query Frontend, this data can be cached either in Memcached, Redis, or in memory.

Index cache

The index cache is used for caching the indices of TSDB blocks, which, in turn, should speed up lookups for postings and series (see *Chapter 3*). The index tends to be the largest individual file within a block, so caching it can reduce time waiting for it to be downloaded and read.

Chunk cache

The chunk cache is for storing actual TSDB chunks and associated metadata without the need to re-fetch it from object storage. This operates independently of the index cache but has a bigger impact on potentially reducing your object storage requests if you're worried about being charged more or rate-limited since there is only one index file per block, whereas there are many chunk segment files. You can use the same cache backend for this as the index cache if you so choose.

Thanos Ruler

Thanos Ruler enables a unique feature for advanced use cases: evaluating Prometheus rules across multiple Prometheus instances. For example, consider a service you have deployed across multiple regions with a Prometheus deployment in each region responsible for monitoring its corresponding instance of the service. Thanos Ruler would enable you to evaluate Prometheus rules across all of those Prometheus instances to obtain a more holistic view of your service. This is great for measuring things such as **service-level objectives** (**SLOs**).

Thanos Ruler accomplishes this by connecting to one or more Thanos Query endpoints to run queries against. If more than one is specified, it performs round-robin balancing of queries. In other words, a rule's query is not evaluated by every specified Query instance – only one is chosen and used per query.

Data produced by evaluating recording rules is stored in a local TSDB in the same manner that it would be on Prometheus. However, Thanos Ruler is not intended to store TSDB blocks locally for long periods. Instead, blocks are uploaded to object storage once they are flushed (every 2 hours; this can be configured via the `--tsdb.block-duration` flag). To account for potential issues uploading to object storage, though, blocks will be retained locally for 48 hours by default (configurable via the `--tsdb.retention` flag).

In addition to recording rules, Thanos Ruler can also be connected to Alertmanager instances to evaluate and send alerting rules. In practice, I've found this to be a less common use case since users are discouraged from relying on Thanos Ruler to replace local rule evaluations from individual Prometheus servers. Thanos Ruler should be *additive* rather than *substitutive*. This is because PromQL expressions in rules are effectively guaranteed to never fail – presuming that your query is valid – when running locally on a Prometheus. In contrast, Thanos Ruler depends on Thanos Query being up and therefore transitively also on Thanos Sidecars and downstream Prometheus instances being reachable.

This is not to say that you *shouldn't* use Thanos Ruler for alerting rules, but I do encourage you to take care when deciding which alerting rules should be added to Thanos Ruler versus individual Prometheus instances.

Stateless mode

While the default mode of operation for Thanos Ruler is stateful – since it stores query results in a local TSDB – you can also run in it stateless mode by providing a remote write configuration via the `--remote-write.config-file` or `--remote-write.config` flag. This follows the same syntax as the Prometheus remote write configurations from *Chapter 9*.

This mode was born out of and is recommended for use cases in which you need to be able to horizontally scale Thanos Ruler with reliable storage and deduplication of resultant time series. In stateless mode, Thanos Ruler will send the time series data it produces to the configured remote storage backend(s). The configured backend can be any backend that accepts the Prometheus remote write protocol, including an actual Prometheus instance or the Thanos Receive component (which is the final component we'll discuss).

When running in this mode, Thanos Ruler will no longer expose a StoreAPI endpoint since it will have no queriable data locally. Consequently, it also will not upload any blocks to object storage – instead, it will use a WAL-only storage based on how Prometheus operates when it is running in Agent mode. However, it will still be capable of sending alerts to Alertmanager instances.

Deploying Thanos Ruler

To deploy Thanos Ruler, we will once again be leveraging manifests that leverage the kube-thanos project built for this book. We'll be using the default `stateful` mode and some basic rules to demonstrate this.

All we need to run is the following `kubectl` command:

```
$ kubectl apply -f https://raw.githubusercontent.com/PacktPublishing/
Mastering-Prometheus/main/ch10/manifests/thanos-ruler.yaml
```

After a short period, we should be able to query the time series created by our rules via Thanos Query. You can once again port-forward to it using the same instructions provided in the *Thanos Query Frontend* section and query for one of the deployed rules, such as `:thanos_objstore_bucket_operation_duration_seconds:histogram_quantile`, to see if it's working! Alternatively, you could query `{rule_replica="thanos-rule-0", __name__=~".+"}` to see all the recording rule results from Thanos Ruler.

Thanos Receiver

Thanos Receiver (or "Receive") is the last of our components to deploy and arguably has the potential to be the most complex Thanos component in your stack, depending on how you configure it. This is mostly because it is extremely configurable for multi-tenant and/or large-scale use cases. However, since it focuses primarily on receiving remote write data, we'll skip diving too deep into the details of remote write since you're already familiar with it from the previous chapter.

Like other Thanos components, Thanos Receive is also intended to connect to object storage to upload TSDB blocks. It maintains a local TSDB while receiving data but will upload blocks to object storage when they are flushed to disk every 2 hours. Unlike Thanos Ruler, it is also intended to maintain a local copy of data for a longer period – by default, 15 days (configurable via the `--tsdb.retention` flag).

> **A note on deduplication**
>
> As opposed to other remote write destinations such as VictoriaMetrics or Mimir, Thanos Receive does not have built-in deduplication capabilities. Instead, to achieve deduplication of data sent to Thanos Receive, consider configuring vertical compaction in the Thanos Compactor component.

To be able to scale Thanos Receive, it implements a **hashring** to efficiently distribute incoming metrics across multiple Receive backends. By default, it utilizes the same simple `hashmod` algorithm we've seen elsewhere. However, this can cause issues when you're scaling your quantity of Thanos Receive instances up or down since time series may stop going to the same backend they were previously routed to. Consequently, the Ketama consistent hashing algorithm is now recommended for use in Thanos Receive; you can do this by adding the `--receive.hashrings-algorithm=ketama` flag to your Thanos Receive deployment. Evaluating the details of consistent hashing algorithms versus more simple algorithms such as `hashmod` is beyond the scope of this book, but suffice it to say that the Ketama algorithm is less disruptive when adding or removing Receive instances from the pool.

Hashrings are defined via a JSON configuration file (specified via `--receive.hashrings-file`) that lists all the members of the configured hashring (you may have multiple hashrings when using multi-tenancy features). The configuration may look something like this:

```
[
    {
        "endpoints": [
            "192.168.1.128:10901",
            "192.168.1.129:10901",
            "192.168.1.130:10901"
        ]
    }
]
```

When running a remote write backend such as Thanos Receive, data durability is of the utmost importance. For this reason, Thanos Receive also allows you to configure a **replication factor** for incoming data. When configuring a replication factor, incoming data must be successfully written to a majority of nodes within the specified replication factor. Therefore, if data is not successfully written to (`REPLICATION_FACTOR + 1`) / 2 nodes, then the ingestion will be considered to have failed. For example, with a replication factor of 3, writes must succeed to at least two Receive backends for the overall request to be considered a success. This can be configured via the `--receive.replication-factor` flag and defaults to 1.

Much of the rest of the configuration for Thanos Receive centers upon supporting multi-tenancy, which – although critically important for niche use cases such as running Thanos as a managed customer service – is not relevant to this book. So, let's skip ahead to the good part and get this final component deployed!

Deploying Thanos Receiver

To deploy Thanos Receiver, we will once again be leveraging manifests that leverage the kube-thanos project built for this book. We'll be using the default stateful mode and some basic rules to demonstrate this.

All we need to run is the following `kubectl` command:

```
$ kubectl apply -f https://raw.githubusercontent.com/PacktPublishing/
Mastering-Prometheus/main/ch10/manifests/thanos-receiver.yaml
```

Then, we can configure our Prometheus instance so that we can send data to it via remote write by adding a remote write configuration pointed at `http://thanos-receive.prometheus.svc.cluster.local:19291/api/v1/receive`. We can accomplish this by applying the relevant Helm values from this book's GitHub repository, like so:

```
$ helm upgrade --namespace prometheus \
    --version 47.0.0 \
    --values mastering-prometheus/ch10/prom-thanos-receiver-values.
yaml \
    mastering-prometheus \
    prometheus-community/kube-prometheus-stack
```

This is redundant and unnecessary in practice since we already have Thanos Sidecar shipping our data to object storage. However, it'll serve our testing purposes.

Once deployed, we can port forward to Thanos Query Frontend once again so that we can see our metrics coming in. We can filter for metrics coming into Thanos Receive by using the automatically attached labels of `receive`, `replica`, and/or `tenant_id`. For example, try using the `up{receive="true"}` PromQL query. Alternatively, you can use Thanos Query's store filtering feature to only query the Thanos Receive endpoint.

Thanos tools

In addition to all of the Thanos components we have reviewed thus far, the Thanos CLI also has a `thanos tools` sub-command. This contains a variety of helpful tools, primarily for interacting with object storage buckets and the TSDB blocks within them. It also contains a command for validating recording and alerting rules used by Thanos Ruler.

Since these tools are primarily for use in existing, established environments, we won't cover them individually in this book. Nevertheless, they may be worth experimenting with before cleaning up the Thanos components you've deployed in this chapter. Within any of the Thanos pods we deployed in this chapter, you can run `thanos tools --help` to see all of the available options.

Cleanup

Now that we're done experimenting with the suite of Thanos components, you can clean up your environment by reverting to our simple Prometheus deployment via Helm, like so:

```
$ helm upgrade --namespace prometheus \
    --version 47.0.0 \
    --values mastering-prometheus/ch10/prom-values.yaml \
    mastering-prometheus \
    prometheus-community/kube-prometheus-stack
```

Additionally, you can delete all of the Thanos components and our object storage configuration via `kubectl`:

```
$ kubectl delete secret thanos-objstore-config
$ kubectl delete -f https://raw.githubusercontent.com/PacktPublishing/
Mastering-Prometheus/main/ch10/manifests/thanos-compact.yaml
$ kubectl delete -f https://raw.githubusercontent.com/PacktPublishing/
Mastering-Prometheus/main/ch10/manifests/thanos-query.yaml
$ kubectl delete -f https://raw.githubusercontent.com/PacktPublishing/
Mastering-Prometheus/main/ch10/manifests/thanos-queryfrontend.yaml
$ kubectl delete -f https://raw.githubusercontent.com/PacktPublishing/
Mastering-Prometheus/main/ch10/manifests/thanos-store.yaml
$ kubectl delete -f https://raw.githubusercontent.com/PacktPublishing/
Mastering-Prometheus/main/ch10/manifests/thanos-ruler.yaml
$ kubectl delete -f https://raw.githubusercontent.com/PacktPublishing/
Mastering-Prometheus/main/ch10/manifests/thanos-receiver.yaml
```

Finally, if you're not using Linode Object Storage for anything else, you can cancel your Object Storage service by following this guide: `https://www.linode.com/docs/products/storage/object-storage/guides/cancel/`.

> **Note**
>
> If you do not cancel the service, you will continue to be charged a small monthly fee, regardless of whether or not you have any buckets in use.

Summary

In this chapter, we went through all of the Thanos components that are available to gain a greater understanding of the comprehensive suite of features offered by the Thanos project.

We learned how Thanos Sidecar enables long-term storage of metrics in object storage and distributed querying through Thanos Query, how Thanos Compactor operates on those uploaded TSDB blocks in object storage to compact and downsample them, and how Thanos Store retrieves them from object storage on-demand for queries.

We saw how Thanos Query enables distributed querying of metrics from the various components that implement Thanos's gRPC StoreAPI, and how Thanos Query Frontend enables more efficient use of Thanos Query instances through caching, query sharding, and splitting.

We utilized Thanos Ruler so that we could evaluate Prometheus alerts and rules across all endpoints connected to a Thanos Query instance.

Finally, we learned about and deployed Thanos Receiver so that we can support remote writing data to Thanos.

In the next chapter, we'll learn about the technology that was used to template and generate all of the Kubernetes manifests to deploy these Thanos components: Jsonnet. We'll explore its extensive use within the Prometheus ecosystem and community, learn some fundamentals about its language, and explore the concept of "mixins," which leverage Jsonnet to provide reusable and customizable Prometheus rules and Grafana dashboards for a variety of commonly deployed applications.

Further reading

To learn more about the topics that were covered in this chapter, take a look at the following resources:

- *Thanos's gRPC StoreAPI*: `https://github.com/thanos-io/thanos/blob/9a5b4fa9c6e789cfc0c3bd648e183a6d4aea4218/pkg/store/storepb/rpc.proto`

- *Thanos Sidecar*: `https://thanos.io/v0.32/components/sidecar.md`

- *Thanos Compactor*: `https://kubernetes.io/docs/tasks/tools/#kubectl`

- *Thanos DNS Discovery*: `https://thanos.io/v0.32/thanos/service-discovery.md/#dns-service-discovery`

- *Thanos Query PromQL Engine*:

  - `https://github.com/thanos-io/thanos/blob/main/docs/proposals-accepted/202301-distributed-query-execution.md`

  - `https://github.com/thanos-io/promql-engine`

- *Thanos Query Frontend*: `https://thanos.io/v0.32/components/query-frontend.md/`

- *Handling instant queries*: `https://github.com/thanos-io/thanos/discussions/6472`

- *Vertical sharding*: `https://thanos.io/v0.32/proposals-accepted/202205-vertical-query-sharding/`

- *Thanos Store external label partitioning*:

    - `https://thanos.io/v0.32/thanos/sharding.md/`

    - `https://github.com/observatorium/observatorium/blob/bf1304b0d7bce2ae3fefa80412bb358f9aa176fb/environments/openshift/manifests/observatorium-template.yaml#L1514-L1521`

- *Thanos Ruler*:

    - `https://thanos.io/v0.32/components/rule.md/`

    - *Stateless mode*: `https://thanos.io/tip/proposals-done/202005-scalable-rule-storage.md/`

- *Thanos Receive*:

    - *Ketama consistent hashing algorithm*: `https://www.metabrew.com/article/libketama-consistent-hashing-algo-memcached-clients`

11

Jsonnet and Monitoring Mixins

Have you ever tried to template YAML? It's a nightmare! Most templating languages like Jinja (used by Ansible and Salt), or even Go's templating language, handle whitespace in generally unintuitive ways to most users. It's devilishly easy to accidentally add extra space into a template's output. Combined with YAML's whitespace sensitivity, it's a recipe for headaches. And yet – like it or not – YAML is the de-facto configuration language of the present time, used for everything from deploying apps to Kubernetes to configuring your Prometheus servers. So, how do we balance the tedious, error-prone management of large YAML files by hand with the arguably more error-prone desire to template our way out of the repetitiveness?

One solution that has seen relatively broad adoption within the Prometheus community is using a language called **Jsonnet** to handle generating YAML files. Jsonnet is a superset of JSON that extends the common data format to provide features such as variables, conditionals, functions, and imports. Leveraging these features, users are able to significantly reduce boilerplate text in their configuration files and easily reuse data throughout and across files. In Prometheus, this may look like templating out the sample label matchers to every PromQL query in a `rules` file or populating the same default settings for each Slack receiver you define in Alertmanager.

The language itself can be a bit intimidating at first, but in the end, you'll be able to have the confidence to know that you're generating valid configuration files using significantly fewer lines of code.

With that in mind, in this chapter, we'll be covering these main topics:

- Overview of Jsonnet
- Using Jsonnet
- Monitoring Mixins

Let's get started!

Technical requirements

For this chapter, we'll be stepping away from our Prometheus deployment for a bit and doing things locally. Consequently, you'll need these tools installed to follow along:

- **jsonnet**: `https://github.com/google/go-jsonnet/blob/fed90cd9cd733a87f9fb27cfb32a3e08a7695603/README.md#installation-instructions`

- **jsonnet-bundler**: `https://github.com/jsonnet-bundler/jsonnet-bundler#install`

- **promtool**: `https://github.com/prometheus/prometheus/releases/tag/v2.46.0`

The code used in this chapter is available at `https://github.com/PacktPublishing/Mastering-Prometheus`.

Overview of Jsonnet

Jsonnet was born out of Google by an engineer who wanted to improve upon and extend an internal configuration language known as **Google Configuration Language** (**GCL**). It was initially released in 2014 – the same year as Kubernetes. It is truly fortuitous that we received a tool that can generate YAML *and* the biggest abuser of YAML in the same year. Google giveth and Google taketh away.

At its core, Jsonnet is an extension of JSON – the common, ubiquitous method for sharing data between applications in a human-readable way. This means that any JSON document is also a valid Jsonnet program and will be emitted unchanged when run. Naturally, requiring Jsonnet to be a language that must operate within the constraints of the JSON specification can be limiting, and you will undoubtedly encounter oddities in the language attributable to those constraints.

Jsonnet does not endeavor to be a fully featured programming language as blissful to use as, say, Python, Go, or your language of choice. However, what it lacks in general-purpose usability, it more than makes up for in its specialized applicability for generating structured data.

With the growing industry-wide focus on **Infrastructure as Code** (**IaC**) and GitOps, it's become more important than ever to be able to manage hundreds of large, sprawling configuration files effectively and efficiently. Doing so by hand is not only tedious but highly error-prone. We see this in Prometheus through not only managing a configuration file that can easily stretch into thousands of lines but also through numerous rules files that tend to consist of many repeated blocks of text (e.g., label matchers in an `expr` key or repeated annotations). With Jsonnet, we can reduce the lines of code we need to manage and the likelihood of errors, such as forgetting to update a label matcher in *all* of our alerting rules for a service. But we've talked enough about what Jsonnet can do for us. Let's see what it actually looks like.

Syntax

Jsonnet's syntax will be familiar to anyone who has ever looked at a JSON document (which should include you unless you picked this book up by accident, in which case, congrats for making it this far). There are a lot of curly braces. There are a lot of nested values. There is a lot of optional whitespace. Everything is an object. But fear not! There are plenty of improvements over standard JSON, including – if you can believe it – *comments*.

> **Important note**
> This chapter is heavy on code snippets, and I highly recommend leveraging the GitHub repository for this book during this chapter rather than trying to transpose each code block to your computer by hand.

As I said before, the simplest Jsonnet program is a valid JSON document. Let's test that out. Put the following in a file we'll call `vanilla.jsonnet`:

```
{
    "book": "Mastering Prometheus",
    "author": {
        "name": "Will Hegedus",
        "job": "SRE Manager",
        "favoriteColor": "orange"
    }
}
```

Now, we can run that file through our `jsonnet` CLI tool, and the output will be exactly the same!

```
$ jsonnet ch11/vanilla.jsonnet
{
    "author": {
        "favoriteColor": "orange",
        "job": "SRE Manager",
        "name": "Will Hegedus"
    },
    "book": "Mastering Prometheus"
}
```

Well… exactly the same in composition, not necessarily in ordering. The astute observer may have noticed that the order of the keys changed in the output since Jsonnet organizes them alphabetically. There is no way around this behavior at the time of writing, but a GitHub issue has been open to add support for custom ordering since 2018: `https://github.com/google/jsonnet/issues/407`. Ordering doesn't matter to computers, but it can certainly aid in human readability. Nevertheless, we soldier on.

> **What's an object?**
>
> **Object** can be a loaded term in tech. For the purposes of this chapter, an object refers to the key-value data structure from JSON and JavaScript. If you're familiar with other programming languages, it's akin to a map in Go, a dictionary in Python, or a HashMap in Java.

While it's a neat party trick that Jsonnet can input and output JSON, we're not using any of the language's features yet! Let's see how we can change that by introducing variables.

Variables

In Jsonnet, variables are declared using the keyword `local` and can be declared inside or outside an object. They can then be directly referenced in an unquoted field. We can variabilize our example in a new file called `variables.jsonnet` like so:

```
local name = "Will Hegedus";
{
    local bookTitle = "Mastering Prometheus",

    "book": bookTitle,
    "author": {
        "name": name,
        "job": "SRE Manager",
        "favoriteColor": "orange"
    }
}
```

Rendering this new file through Jsonnet, we'll get the same output as before:

```
$ jsonnet ch11/vanilla.jsonnet
{
    "author": {
        "favoriteColor": "orange",
        "job": "SRE Manager",
        "name": "Will Hegedus"
    },
    "book": "Mastering Prometheus"
}
```

However, there's not much point in using variables if you only use them in one place. Let's see how variables might look in a set of Prometheus alerting rules by defining a variable for a common set of label matchers we want to use in our alerts. In a file called `variables_prometheus.jsonnet`, put the following code:

```
{
    groups: [
        {
            local matchers =
'service="prometheus",environment="prod"',
            name: "chapter11",
            rules: [
                {
                    alert: "SSHDown",
                    expr: 'last_over_time(probe_success{job="blackbox_
ssh",%s}[5m]) != 1' % matchers,
                    "for": "5m",
                },
                {
                    alert: "ICMPDown",
                    expr: 'last_over_time(probe_success{job="blackbox_
icmp",%s}[5m]) != 1' % matchers,
                    "for": "5m",
                }
            ]
        }
    ]
}
```

Using this code, we template out two alerting rules that include the same common label matchers and the specific one for the alert. It should look like this:

```
$ jsonnet ch11/variables_prometheus.jsonnet

{
    "groups": [
        {
            "name": "chapter11",
            "rules": [
                {
                    "alert": "SSHDown",
                    "expr": "last_over_time(probe_success{job=\"blackbox_
ssh\",service=\"prometheus\",environment=\"prod\"}[5m]) != 1",
                    "for": "5m"
                },
```

```
                {
                    "alert": "ICMPDown",
                    "expr": "last_over_time(probe_success{job=\"blackbox_
   icmp\",service=\"prometheus\",environment=\"prod\"}[5m]) != 1",
                    "for": "5m"
                }
            ]
        }
    ]
}
```

Note that in our `variables_prometheus.jsonnet` file, we surrounded the `for` key with quotation marks. This is because `for` has a special meaning in Jsonnet to perform list and object comprehensions, which we'll cover more later.

We also just got our first glimpse at string interpolation in Jsonnet. We used `%s` as a placeholder in the `expr` strings and told Jsonnet to substitute it with the value of the `matchers` variable. This syntax may look familiar to those with experience in Python, C, Go, or other languages with `printf`-style formatting, but let's take a deeper look at it.

String interpolation

String interpolation – also known as variable substitution or variable expansion – is a method where a string can be defined with placeholders that are dynamically substituted with variables. This is distinct from string *concatenation*, in which you combine multiple strings using an addition (+) operator. The difference is shown in this example:

```
{
    local program = "Jsonnet",
    interpolation: "I am using %s to interpolate this string." %
program,
    concatenation: "I am using " + program + " to concatenate this
string."
}
```

Generally, interpolation tends to be easier to write and read, especially when substituting multiple variables. Additionally, interpolation offers added formatting functionality during substitution. You can use most of the same variable type specifiers as languages with `printf` formatting. A non-exhaustive list of some of the most common is outlined in the table below.

Placeholder	Meaning	Output
%s	String	"12"
%d	Digit (integer)	12
%f	Floating point number	12.000000

Placeholder	Meaning	Output
%.2f	Floating point limited to 2 decimal places	12.00
%x	Hexadecimal number	c
%o	Octal number	14

Table 11.1 – printf-style substitutions

Variables are substituted in order of appearance from left to right. So, the first %s (or %d, or another placeholder) in a string would be replaced with the first variable following the modulo operator (%) outside the string, and so on. As you might expect, strings with many substitutions can be quite cumbersome to read, write, and maintain since you need to make sure your ordering of substitutions remains accurate. Conveniently, there is another syntax available to us to simplify this.

By combining the above placeholder syntax with parenthetical variables, we can provide an *object* as the argument to the modulo operator, and its keys can be used in-line for substitutions. Continuing our previous example, we'll use a common pattern of defining an object (we'll call it _config) with our variables:

```
{
    local _config = {
        program: "Jsonnet",
        os: "Linux"
    },
    inlineInterpolation: "I am using %(program)s on %(os)s to
interpolate this string." % _config,
}
```

In my opinion, this is the most readable syntax since you clearly see which variables are being substituted into which part of the string. You still leverage the placeholders like %s, but just put the variable between the modulo and the letter that specifies the substitution's type.

This comes in handy for exceptionally long strings, which are also supported by Jsonnet using the special ||| characters to denote a block of text. These are used in place of quotation marks as delimiters for the start and end of the text and are helpful for long strings such as PromQL queries in your rules files. This allows for more nicely formatted PromQL expressions to still be used. For example, we can do substitution in a rule definition like this:

```
{
    groups: [
        {
            local _config = {matchers:
'service="prometheus",environment="prod"'},
            name: "chapter11",
```

```
        rules: [
            {
                alert: "SSHDown",
                expr: |||
                    last_over_time(
                        probe_success{job="blackbox_
ssh",%(matchers)s} [5m]
                    ) != 1
                ||| % _config,
                "for": "5m",
            },
        ]
    }
  ]
}
```

Once we get into defining a configuration variable, though, we are limited in its usefulness if we combine multiple files the way that Jsonnet is intended. This is because the variables are scoped locally to the file in which they are defined – they cannot be overridden. There are ways around this by using external variables or top-level arguments, but these also become verbose, with the need to pass along a bunch of command-line flags when executing your Jsonnet files. Consequently, we will skip them in favor of the concept of "mixins" within Jsonnet.

> **Mixology**
>
> The concept of mixins is not unique to Jsonnet but rather an inheritance mechanism in some object-oriented program languages. In this compositional inheritance model, objects are provided access to variables and methods associated with the mixin class without having to be a child of that class. In this way, they are considered "included" rather than "inherited" from. If you're interested in learning more, the concept is the subject of an academic paper from Gilad Bracha and William Cook, available online at https://www.bracha.org/oopsla90.pdf.

With mixins, we leverage *hidden* fields to provide configuration, which we'll explore next.

Hidden fields

Hidden fields are a unique concept within Jsonnet in which you can define fields in an object that is *not* included in the output of the file. One common use of these fields is to provide a way of configuring the templating of a JSON object within the object itself. This ties in heavily with the idea of imports, which we'll discuss later.

Hidden fields are defined using two colons (: :) instead of one as the delimiter between a key and a value. You can see this in action via the following Jsonnet code:

```
{
    invisibleField:: "You can't see me!",
    normalField: "Boo!"
}
```

When rendered, it provides this output:

```
$ jsonnet ch11/hiddenFields.jsonnet
```

```
{
    "normalField": "Boo!"
}
```

We can extend our previous example of templating a Prometheus rule using a hidden `config` field instead of a local variable like so:

```
{
    _config:: {
        matchers: 'service="prometheus",environment="prod"',
        forDuration: '5m'
        },
    groups: [
        {
            name: "chapter11",
            rules: [
                {
                    alert: "SSHDown",
                    expr: |||
                        last_over_time(
                            probe_success{job="blackbox_
ssh",%(matchers)s}[5m]
                        ) != 1
                    ||| % $._config,
                    "for": $._config.forDuration,
                },
            ]
        }
    ]
}
```

Note that how we provide `config` to the `modulo` operator has now changed. We use a dollar sign ($) to refer to the object containing the `_config` field. This is one of the ways that Jsonnet allows you to refer to other objects within a Jsonnet program. Let's take a look at those options.

Object references and inheritance

Since JSON – and, by extension, Jsonnet – revolves around the core data structure of an object, which itself often contains nested objects, it is critical functionality for Jsonnet to provide a way to reference those objects. There are three key ways to refer to objects in a nested structure in Jsonnet: `super`, `self`, and $.

If you have experience with some object-oriented programming languages, `super` and `self` may seem familiar. The `self` keyword is a reference to the object in which the reference is occurring. The `super` keyword is a reference to the parent of the object in which the reference is occurring. This is only applicable if you leverage object inheritance by combining two objects and is *not* a way to reference a field in the object one level above a nested object.

Lastly – and perhaps most uniquely – the $ symbol is a short-hand way to reference the top-most object of the object in which the reference is occurring. If all this nesting is challenging to wrap your head around, let's see what it looks like in action instead of just talking about it. The following Jsonnet code snippet shows an example of all three in action:

```
{
    local parentObject = {
        parentField:: "Hello, from the parent!"
    },

    topField:: "Hello, from the top!",

    child: parentObject + {
        childField:: "Hello from the child!",
        a: $.topField,
        b: super.parentField,
        c: self.childField
    }
}
```

Since we hid the actual fields using double colons, they should only be visible where we reference them within the `child` object. A quick test to confirm provides the following output:

```
$ jsonnet ch11/references.jsonnet

{
    "child": {
        "a": "Hello, from the top!",
```

```
        "b": "Hello, from the parent!",
        "c": "Hello from the child!"
    }
}
```

In our code snippet, the child object is an extension of `parentObject` with some additional fields. This extension uses the simple addition (+) operator and is how **object inheritance** is performed in Jsonnet.

We extend `parentObject` by adding some additional fields, but `parentObject` becomes the "parent" object of `child` when we perform this combination (you can see how clever I am with my naming). Since we're extending parentObject, we're also inheriting its fields. Consequently, `child` also receives the hidden `parentField` key. We can confirm this by using a standard library function (`std.objectFieldsAll`) that shows all fields of an object.

> **Jsonnet's standard library**
>
> Like any other programming language, Jsonnet provides a "standard library" of commonly used functions available in all files. Throughout the rest of this chapter, we'll use several standard library functions for simplicity. They are always prefixed with `std.` and their names should be self-explanatory when you see them. For a list of all standard library functions – including explanations of what each does – consult the Jsonnet project's documentation at `https://jsonnet.org/ref/stdlib.html`.

Extending the previous example, our code now looks like this:

```
{
    local parentObject = {
        parentField:: "Hello, from the parent!"
    },

    topField:: "Hello, from the top!",

    child: parentObject + {
        childField:: "Hello from the child!",
        a: $.topField,
        b: super.parentField,
        c: self.childField,
        d: std.objectFieldsAll(self)
    }
}
```

The output now looks like this:

```
$ jsonnet ch11/inheritance.jsonnet

{
    "child": {
        "a": "Hello, from the top!",
        "b": "Hello, from the parent!",
        "c": "Hello from the child!",
        "d": [
            "a",
            "b",
            "c",
            "childField",
            "d",
            "parentField"
        ]
    }
}
```

Sure enough, there is `parentField` included as one of the fields that `child` contains.

We can greatly reduce the need for duplicated code through references and inheritance and keep our templates **DRY** (standing for **Don't Repeat Yourself**). The other way to keep our templates DRY is by using imports and overriding deeply nested fields.

Imports and overrides

Jsonnet is intended to be used as a way to simplify configuration and data structures. One of the most straightforward ways to keep large configurations simple and manageable is by breaking them into smaller pieces. In Prometheus, this may look like having a separate Jsonnet file for each of your rule groups that all get combined. The way that Jsonnet is able to accomplish this is through the concept of **imports**.

Imports allow you to either take the content of the imported file and use it directly or – alternatively – modify it in some way. By convention (i.e., it's not a *requirement*), files written for the primary purpose of being imported should use the `.libsonnet` file extension instead of `.jsonnet`. This is to distinguish the fact that the file is intended to be used as a library, akin to how other programming languages are able to import and use other libraries/modules.

There are two main ways that Jsonnet goes about imports: *direct imports* and *variablized imports*.

Direct imports take the content of a file and directly import it – it is included as is in the file it is imported into. To demonstrate this, we'll make two files – one to define an object and the other to import it. We'll call one `importer.jsonnet` and the other `imported.libsonnet`.

In `imported.libsonnet`, we'll put our actual data:

```
{
    filename: "imported.libsonnet",
    monitoringStack: [
        "prometheus",
        "grafana"
    ]
}
```

Then, to use it, all we need to add to `importer.jsonnet` is this:

```
import 'imported.libsonnet'
```

Now, we can run the Jsonnet CLI to render `importer.jsonnet` and we'll see the content of `imported.libsonnet` within it:

```
$ jsonnet ch11/importer.jsonnet

{
    "filename": "imported.libsonnet",
    "monitoringStack": [
        "prometheus",
        "grafana"
    ]
}
```

We can even include multiple imports that are combined into the same output. I'll import the `vanilla.jsonnet` file from the beginning of the chapter, too. Now my `importer.jsonnet` file looks like this:

```
(import 'imported.libsonnet')
+ (import 'vanilla.jsonnet')
```

And my output looks like this:

```
$ jsonnet ch11/importer.jsonnet

{
    "author": {
        "favoriteColor": "orange",
        "job": "SRE Manager",
        "name": "Will Hegedus"
    },
    "book": "Mastering Prometheus",
```

```
    "filename": "imported.libsonnet",
    "monitoringStack": [
        "prometheus",
        "grafana"
    ]
}
```

This allows us to combine the contents of multiple jsonnet/libsonnet files into one resulting document, which is great. But what if we want to modify the contents of the files we're importing? After all, if they're intended to be used as libraries, they're most likely supposed to be modified in some way. To do this, we'll use variablized imports.

Variablized imports are when we assign the content of the file we're importing to a variable so that we can modify it before exporting it. Let's create a new file called var_importer.jsonnet where we'll add Thanos to the list of things in our monitoring stack. In order to add it while preserving the existing contents, we'll use the special +: field syntax. This syntax will override the field to merge our changes with the existing value for the field rather than completely overwrite it. This behavior is possible with both visible fields (+:) and with hidden fields (+::).

To accomplish this, the contents of var_importer.jsonnet will look like this:

```
local i = import 'imported.libsonnet';

i {
    monitoringStack+: ['thanos']
}
```

Now, when we render var_importer.jsonnet, the output will look similar to importer. jsonnet's initial output but with thanos included:

```
$ jsonnet ch11/var_importer.jsonnet

{
    "filename": "imported.libsonnet",
    "monitoringStack": [
        "prometheus",
        "grafana",
        "thanos"
    ]
}
```

Pretty useful, right? Understanding how imports and overrides work will be foundational for using Monitoring Mixins, which we'll explore later in this chapter. For now, though, let's keep trekking along to see how we can simplify the creation of lists and objects through comprehensions.

List and object comprehensions

Comprehensions are a syntactic feature of several programming languages, including – notably – Python. They greatly simplify the definition of things like lists or objects by being able to concisely iterate through other values to populate the data structure (as opposed to using a traditional `for` loop).

List comprehensions are straightforward and follow the following syntax:

```
$expr for $variable in $iterable
```

Optionally, you can add an `if` condition to control whether or not `$expr` should be executed. In that case, the syntax looks like this:

```
$expr for $variable in $iterable if $condition
```

Let's see what that looks like in practice, though. In a file named `comprehensions.jsonnet`, we'll create a hidden field with a list of foods and an exposed field that uses a list comprehension to only shows fruits:

```
{
    foods:: [
        {name: "apple", kind: "fruit"},
        {name: "broccoli", kind: "vegetable"},
        {name: "banana", kind: "fruit"},
        {name: "carrot", kind: "vegetable"}
    ],

    fruits: [f.name for f in self.foods if f.kind != "vegetable"]
}
```

Running that file through the Jsonnet interpreter, our output looks like this:

```
$ jsonnet ch11/comprehensions.jsonnet
{
    "fruits": [
        "apple",
        "banana"
    ]
}
```

Huzzah! No vegetables! List comprehensions make it much easier to populate sets of data and even filter them. However, we may want to be able to create objects the same way. The syntax is pretty similar, but we need to take into account the concept of **computed field names** now.

A computed field name means that the name of a field in an object is unknown until it is computed (for example, in an object comprehension). To signify this in Jsonnet, we must wrap the field name in brackets ([ ]). We'll do this to create a grocery list in our comprehensions.jsonnet file. Continuing our faux distaste for vegetables, we'll conditionally add a quantity of 1 for vegetables but 5 for fruits. To accomplish this, our file looks like this:

```
{
    foods:: [
        {name: "apple", kind: "fruit"},
        {name: "broccoli", kind: "vegetable"},
        {name: "banana", kind: "fruit"},
        {name: "carrot", kind: "vegetable"}
    ],

    fruits: [f.name for f in self.foods if f.kind != "vegetable"],
    groceryList: {
        [f.name]: {qty: if f.kind == "fruit" then 5 else 1}
        for f in self.foods
    }
}
```

The output of the rendered file then looks like this:

```
$ jsonnet ch11/comprehensions.jsonnet

{
   "fruits": [
      "apple",
      "banana"
   ],
   "groceryList": {
      "apple": {
         "qty": 5
      },
      "banana": {
         "qty": 5
      },
      "broccoli": {
         "qty": 1
      },
      "carrot": {
         "qty": 1
      }
   }
}
```

The conditional syntax that we used is new to what we've done thus far, but it should be straightforward to reason about. This if/then/else syntax can be used to conditionally set values, and the else statements can even be chained with more if statements, like if x > 0 then "positive" else if x < 0 then "negative" else "zero". The else clause is optional. When absent, the value will default to null if the if condition is not met.

With conditionals and comprehensions now understood, we're nearing the end of our tour through most of Jsonnet's syntax. But what is a programming language without functions? You'd be forgiven if you thought that – just because we haven't talked about them yet – Jsonnet can't define functions, but it does! Let's see how they work.

Functions

Functions in Jsonnet are typically defined using the same local keyword as variables and use Pythonic parameter syntax, which supports positional arguments, named arguments, and default values for named arguments. Rather than use a return keyword to control the data returned by a function, functions instead return whatever data is emitted within the function.

Let's look at an example where we define a function to test if a word is a palindrome (a word that is the same both forward and backward). To accomplish this, we'll split the word into an array of individual characters, reverse it, and compare the original array to the reversed array. The code looks like this:

```
local IsPalindrome(word) =
    local chars = std.stringChars(word);
    local reversedChars = std.reverse(chars);
    chars == reversedChars;


{
    level: IsPalindrome("level"),
    prometheus: IsPalindrome("prometheus"),
    racecar: IsPalindrome("racecar"),
    thanos: IsPalindrome("thanos")
}
```

Running this through the Jsonnet interpreter, we get the following output:

```
$ jsonnet ch11/functions.jsonnet

{
   "level": true,
   "prometheus": false,
   "racecar": true,
   "thanos": false
}
```

Notably, the function definition does not have any explicit encapsulation the way many programming languages do (e.g., Go or Java, where the function's body is contained within curly braces). Instead, a semicolon is the delimiter for the end of the definition, like other `local` functions. If the `local` function is defined within an object, the delimiter would be the typical comma.

Functions may also be defined within an object using the `function` keyword. However, functions defined this way *must* be contained within a hidden field since a function definition cannot be manifested as JSON. A simple example of squaring a number would be the following:

```
{
    square:: function(x) x*x,
    foo: self.square(7)
}
```

The output would then simply be this:

```
{
    foo: 49
}
```

Since there are multiple ways to define and use functions, think carefully when defining them and how they should be scoped. For example, if you want your function to be able to be used as a part of a library, it shouldn't be assigned in a `local` definition. If you're defining a function as part of a library, though, you likely need a way to validate the arguments to the function to ensure they are the correct type and/or valid options. How can you put safeguards in place to do such a thing? Well, our final Jsonnet syntax topic will cover just how to do that through the usage of errors.

Errors

Errors are a core component of any programming language, and Jsonnet is no different. Errors allow you to ensure that your templating experience is safe and predictable by enabling data validation and enforcing that specific fields in an object from a library must be overridden.

The language itself can raise errors. For example, looking back to the previous section's function definition example, Jsonnet raises an error if the field containing the function definition is visible. That error looks like this:

```
$ jsonnet ch11/functions.jsonnet

RUNTIME ERROR: couldn't manifest function as JSON
        Field "square"
        During manifestation
```

We're also able to define our own errors, though. This is primarily accomplished through two methods: *explicit errors* and *assertion errors*.

Explicit errors are errors that are explicitly thrown using the `error` keyword followed by a custom error message. One convenient use for this is to force an object field to be overridden by making its default value an error. If it is not overridden, the error is triggered. For example, in a file called `errors.jsonnet`, let's define a base class for a Prometheus rule group. It will have a name field that must be overridden when the base class is used, or an error will be triggered. The code looks like this:

```
local prometheusRuleGroup = {
    name: error "Must provide a name for the Prometheus rule group.",
    rules: []
};


{
    groups: [prometheusRuleGroup],
}
```

Since we use the base object in the `groups` key without overriding the name field, we'll see an error when running this Jsonnet file:

```
$ jsonnet ch11/errors.jsonnet

RUNTIME ERROR: Must provide a name for the Prometheus rule group.
        ch11/errors.jsonnet:2:11-69    object <prometheusRuleGroup>
        Field "name"
        Array element 0
        Field "groups"
        During manifestation
```

To remedy this error, we need to override that name field. Updating our code to do so, our `errors.jsonnet` file should now look like this:

```
local prometheusRuleGroup = {
    name: error "Must provide a name for the Prometheus rule group.",
    rules: []
};


{
    groups: [
        prometheusRuleGroup + {
                name: "general"
            }
        ],
}
```

Running that through Jsonnet, we no longer get an error:

```
$ jsonnet ch11/errors.jsonnet

{
    "groups": [
        {
            "name": "general",
            "rules": [ ]
        }
    ]
}
```

Assertion errors are thrown using the `assert` keyword followed by a conditional expression that must evaluate to `true`, or else an error is thrown. They can be useful for input validation. Optionally, assertion errors can have a custom error message by using the `assert $condition : "$custom_error_message"` syntax. I highly recommend providing these custom error messages to ensure clarity about why Jsonnet is failing to render.

Continuing our example from above, we can use `assert` to add additional validation to enforce that the `rules` field is an array and not empty. In a new file called `assert.jsonnet`, place the following code:

```
local prometheusRuleGroup = {
    name: error "Must provide a name for the Prometheus rule group.",
    rules: []
    assert std.type(self.rules) == "array" : "The rules field must be
an array",
    assert std.length(self.rules) > 0 : "Why are you trying to define
a rule group without any rules?",
};

{
    groups: [
        prometheusRuleGroup + {
            name: "general",
        }
    ],
}
```

Since we do not override the `rules` field in our object currently, we'll see an error when trying to run this code:

```
$ jsonnet ch11/assert.jsonnet
```

```
RUNTIME ERROR: Why are you trying to define a rule group without any
rules?
        ch11/assert.jsonnet:5:5-103
        Checking object assertions
        Array element 0
        Field "groups"
        During manifestation
```

To fix this, we'll need to override the `rules` field so that our code now looks like this:

```
local prometheusRuleGroup = {
    name: error "Must provide a name for the Prometheus rule group.",
    rules: []
    assert std.type(self.rules) == "array" : "The rules field must be
an array",
    assert std.length(self.rules) > 0 : "Why are you trying to define
a rule group without any rules?",
};


{
    groups: [
        prometheusRuleGroup + {
            name: "general",
            rules: [
                {
                    alert: "ScrapeJobDown",
                    expr: "up != 1"
                },
            ]
        }
    ],
}
```

Now, the code will work and provide the following output:

```
$ jsonnet ch11/assert.jsonnet

{
    "groups": [
        {
            "name": "general",
            "rules": [
                {
                    "alert": "ScrapeJobDown",
                    "expr": "up != 1"
```

```
                }
            ]
        }
    ]
}
```

Asserts and errors are powerful tools for enforcing conventions and validating data. However, I must offer a word of caution: Jsonnet is not primarily designed to be a data validation language. Consequently, the stricter that you attempt to make your validation, the more convoluted your code will become, and the difficulty of debugging issues will only increase. For example, trying to iterate over every rule in the `rules` array to ensure it contains specific fields can certainly be done, but it gets messy very fast. Instead, leave detailed validation of the output to external tools such as `promtool` (such as via `promtool check rules`). If you truly need more extensive data validation as it's being templated, I recommend checking out other languages such as CUE (`https://cuelang.org/`) or Nickel (`https://nickel-lang.org/`).

This concludes our tour of Jsonnet's syntax. We covered everything you need to start with Jsonnet and begin using it productively. Nevertheless, we did not cover every nook and cranny of the language. If you're interested in learning more about the language and its specification, I recommend exploring the official project documentation at `https://jsonnet.org/`.

We've explored how to generate data with Jsonnet, but we still need to put it into practice by exploring what it looks like to use Jsonnet in production – especially with regard to Prometheus.

Using Jsonnet

By now, we've harnessed the power and potential of Jsonnet. But what does it really look like to use it outside of the narrowly-scoped, contrived examples we've looked at thus far? How do I use it with Prometheus? This is a Prometheus book, after all. Well, firstly, let's look at how to use Jsonnet to actually generate files instead of just writing to our terminal.

Generating files

In its most simple form, writing to files with Jsonnet is as easy as providing an additional flag (`-o`) and file path to the `jsonnet` CLI. For example, we can write the output of our earlier `variables_prometheus.jsonnet` file to a new file called `out/rules.json` by running the following:

```
$ jsonnet -co ch11/out/rules.json ch11/variables_prometheus.jsonnet
```

The `-c` flag is used in addition to `-o` since `-c` will create parent directories (such as `ch11/out/`) for the output file if they don't exist already. Our `ch11/out/rules.json` file now contains the expected content, but it's still in a JSON format. Is that OK?

Short answer: yes. You can confirm using `promtool` that the rules are valid for Prometheus:

```
$ promtool check rules ch11/out/rules.json

Checking ch11/out/rules.json
  SUCCESS: 2 rules found
```

Long answer: YAML is actually a superset of JSON. Similar to Jsonnet, this means that any valid JSON file *should* be a valid YAML file. There are some caveats to this with typical JSON files. However, there *should* be no odd behaviors when using the JSON output generated by Jsonnet in place of a YAML document.

Nevertheless, it's likely to confuse people if they open a Prometheus rules file expecting YAML and see JSON instead. Conveniently, Jsonnet includes standard library functions to manifest Jsonnet output as something that resembles a more traditional YAML format.

Outputting to YAML

In order to control the format of the output of a Jsonnet file, we need to alter the file slightly. Rather than a command-line flag, we use a standard library function to change the output to YAML.

> **Jsonnet output formats**
>
> In addition to YAML, Jsonnet provides standard library functions for outputting data in TOML, INI, Python, and XML formats.

Jsonnet has two standard library functions that can be used to output data in a YAML format: `std.manifestYamlDoc()` and `std.manifestYamlStream()`. The difference between the two is that `std.manifestYamlDoc()` is for manifesting a single YAML document, whereas `std.manifestYamlStream()` is for writing out multiple YAML documents – separated by the YAML document delimiter of `---` on a line – to a single file.

In order to output our rules file as YAML instead, we'll create a new Jsonnet file called `yaml_rules.jsonnet` that imports our `variables_prometheus.jsonnet` file and wraps it in a call to `std.manifestYamlDoc()`. The code is just one line and looks like this:

```
std.manifestYamlDoc((import 'variables_prometheus.jsonnet'))
```

When rendering the output now, Jsonnet will give us a string instead of an object as before. This means we get a nearly unreadable mess full of newline characters (`\n`) unless we pass the `-S` flag to Jsonnet to tell it to expect the output to be a string. Consequently, to have it look as expected when we render it to a file (`rules.yaml`), our command will look like this:

```
$ jsonnet -So ch11/out/rules.yaml ch11/yaml_rules.jsonnet
```

After running that command, the content of `rules.yaml` should look similar to this:

```
"groups":
- "name": "chapter11"
  "rules":
  - "alert": "SSHDown"
    "expr": "last_over_time(probe_success{job=\"blackbox_
ssh\",service=\"prometheus\",environment=\"prod\"}[5m]) != 1"
    "for": "5m"
  - "alert": "ICMPDown"
    "expr": "last_over_time(probe_success{job=\"blackbox_
icmp\",service=\"prometheus\",environment=\"prod\"}[5m]) != 1"
    "for": "5m"
```

That looks *almost* as expected, but why are the keys quoted? Quoting the keys in YAML output is the default behavior of Jsonnet. It is intended to safeguard against any YAML idiosyncrasies that could crop up when the data is translated to YAML. If it bothers you, the behavior can be disabled by passing `quote_keys=false` as an argument to the `std.manifestYamlDoc()` function.

Outputting to multiple files

As your usage of Jsonnet progresses, you will likely begin to have multiple output files you wish to render. After all, putting all of your Prometheus rules into one big file isn't particularly readable to end users. Rather than engaging in the tedious task of rendering these files one at a time through multiple Jsonnet commands, Jsonnet has yet another feature that will allow us to render multiple output files at once. However, it will once again require some slight modifications to our Jsonnet files and a new command-line flag.

In order to render multiple files, your Jsonnet file should be structured in such a way that it is a single object where the keys correspond to filenames and the values correspond to the contents of that file. In a new `multi_out.jsonnet` file, we'll put the following code:

```
{
    'out/rulesA.yaml': std.manifestYamlDoc(
        (import 'variables_prometheus.jsonnet')
    ),
    'out/rulesB.yaml': std.manifestYamlDoc(
        (import 'assert.jsonnet')
    )
}
```

Then, we'll pass the `-m` flag to Jsonnet when rendering to tell it to use its "multi" mode and output multiple files using the provided directory as the base for file paths:

```
$ jsonnet -Sm ch11/ ch11/multi_out.jsonnet
```

```
ch11/out/rulesA.yaml
ch11/out/rulesB.yaml
```

The content of `rulesA.yaml` should be identical to the `rules.yaml` file we previously created, and `rulesB.yaml` should contain the Prometheus rule from `assert.jsonnet`.

It's all coming together now, huh? You'll be a pro at this in no time. But before we move on to talking about Monitoring Mixins for Prometheus, let's take a look at some nice Jsonnet ecosystem tools to ensure our Jsonnet is uniform and correct.

Formatting and linting

Although Jsonnet adheres to the JSON philosophy of being liberal with formatting style, it does provide an optional tool called `jsonnetfmt` to reformat Jsonnet code to adhere to opinionated style guidelines. Most stylistic choices are configurable via command-line flags (e.g., whether to use `#` or `//` for prefixing comments).

When starting out with Jsonnet, it's perfectly fine to skip using this tool. However, as you develop your own style for writing Jsonnet and begin sharing your Jsonnet code with others to contribute, consider configuring and using `jsonnetfmt` to enforce a consistent style across all your files.

In addition to enforcing syntax, the Jsonnet project provides another tool aimed at catching common issues in Jsonnet files. The `jsonnet-lint` tool can be used as a linter to catch potential syntax issues and edge cases before they produce actual errors during rendering. Like `jsonnetfmt`, it's entirely optional to use, and just because `jsonnet-lint` says there is a potential problem with a file, it does not mean that the file will not render. Nevertheless, I still highly recommend integrating it into your workflow when using Jsonnet.

There are a variety of other tools within the Jsonnet ecosystem (including a Visual Studio Code plugin and language server, both maintained by Grafana Labs), but to talk about them all would be enough content for a whole other chapter. Instead, let us rejoice, for we have come to the culmination of our labors. With our foundational knowledge firmly established, we are finally able to explore the topic of Monitoring Mixins.

Monitoring Mixins

The Monitoring Mixins project was started in 2018 and initially driven primarily by two pillars of the Prometheus community, Tom Wilkie (CTO of Grafana Labs) and Frederic Branczyk (of Prometheus

Operator fame, among many other things). The stated aim of the project is to "define a minimal standard for how to package together Prometheus alerts, Prometheus recording rules, and Grafana dashboards."

The project aims to address the needs of two distinct user bases: consumers and providers. In this context, providers are developers and maintainers of software who wish to be able to provide – in a consistent way – a base-level package of configurable Prometheus alerts, recording rules, and Grafana dashboards for their software. This is in line with the thinking that developers of software are the most equipped to be able to identify what metrics their software exposes and which should be of the most interest to its users. On the opposite side, consumers of mixins want to have a consistent, predictable way to consume and configure these packages across multiple services. Thus, the Monitoring Mixins project was born to service these needs by defining that common framework.

Since its inception, the Monitoring Mixins project has seen a respectable amount of adoption. Numerous open source projects now provide mixins for their software, including projects such as Ceph, Thanos, and Redis. In fact, if you've been following along with this book, you're already using a mixin! The kube-prometheus Helm chart we've been using to deploy our Prometheus environment includes the Kubernetes mixin by default (`https://github.com/kubernetes-monitoring/kubernetes-mixin`).

However, therein lies the problem with mixins thus far: a broad swath of their user base does not even know they are using them. Consumers like you and I cannot extract their full value without knowing what they are or how to configure them. At best, this leaves us with boilerplate Prometheus rules, but at worst, it leaves us with dashboards and alerts that don't even work.

Part of this problem stems from the general lack of familiarity with Jsonnet within the industry. Even if people have heard of it, the resources for learning it are few and far between (something this chapter aspires to help with). Still, I strongly believe in the value provided both by learning Jsonnet in general and by using Monitoring Mixins. If you've made it this far, I hope you can see the potential of this idea, too. With that in mind, let's see what the structure of a monitoring mixin actually looks like.

Mixin structure

At its core, a mixin is a Jsonnet package containing configurable Prometheus rules and/or Grafana dashboards. The recommended schema for a mixin is fairly simple and consists of only four fields.

Field name	Description
`prometheusRules`	Array of Prometheus recording rule groups
`prometheusAlerts`	Array of Prometheus alerting rule groups
`grafanaDashboards`	A dictionary (object) mapping dashboard filenames to dashboard JSON definitions
`_config`	Floating point limited to 2 decimal places

Table 11.2 – Monitoring Mixin fields

Each of these fields is generally a hidden field, and it is the responsibility of the consumer of the mixin to render them out as valid JSON or YAML files. Put together, the structure of a mixin generally looks like this:

```
{
    _config+:: {
        selector: 'job="job_name"',
    },
    grafanaDashboards+:: {
        "dashboard-name.json": {...},
    },
    prometheusAlerts+:: [...],
    prometheusRules+:: [...],
}
```

Many projects that provide mixins will also provide the rendered output using default values. However, if you're going to use a mixin, I highly recommend making the initial commitment to render them yourself. It'll save you time in the long run when you inevitably need to make changes. How do you go about actually using these mixins, though? If they come from some GitHub repo, how am I supposed to use them as a library? Let's go through an example of using the Thanos mixin to see what the process looks like in practice.

Using and extending mixins

For us to use a mixin, we must first be able to import it. However, we must have a copy of it locally to import it. Unfortunately for us, Jsonnet does not have its own package manager. Instead, it's left up to the community to provide that functionality, but thankfully, it has not (yet) succumbed to the package management hellscape of programming languages such as Python (https://xkcd.com/927/). To my knowledge, the primary (only?) Jsonnet package manager is jsonnet-bundler (jb) and was created by Frederic, one of the original authors of the Monitoring Mixins specification.

The jb tool works by downloading a ZIP file of a repository containing mixin code and placing it into a vendor/ directory. It will also automatically manage a dependency file and associated lock file to ensure that other users of your Jsonnet code are able to use the same versions of dependencies.

Presuming you've already installed jb using the installation instructions linked in the *Technical requirements* section of this chapter, we can proceed with creating our dependency file and downloading the Thanos mixin using these commands:

```
$ jb init
$ jb install github.com/thanos-io/thanos/mixin@main
```

You should notice that, when installing the Thanos mixin, jb will also download the dependencies of the Thanos mixin. Therefore, we have everything we need to get started.

To begin using the mixin, we'll need to create a new Jsonnet file (`thanos.jsonnet`) that will import it and override some configuration values. Most mixins will generally contain a `config.libsonnet` file where you can easily see configurable fields. This is a good place to start by seeing what settings you can tweak, but I also tend to poke around in the `vendor/` directory to see what exactly these settings are doing. For example, you can ascertain from looking at the Thanos mixin code that it has a top-level field for configuring each Thanos component, and you can cause the mixin to skip outputting data for a component by setting its field value to `null`.

Since we're unlikely to deploy every Thanos component, let's imagine a simple deployment in which we only run Thanos Sidecar and Thanos Query. We'll define our `thanos.jsonnet` file to look like this in order to only get output related to those components:

```
local thanos = (import 'github.com/thanos-io/thanos/mixin/mixin.
libsonnet') {
    query+:: {
        selector: 'job=~".*thanos-query.*"',
        title: '%(prefix)sQuery' % $.dashboard.prefix,
    },
    sidecar+:: {
        selector: 'job=~".*thanos-sidecar.*"',
        thanosPrometheusCommonDimensions: 'namespace, pod',
        title: '%(prefix)sSidecar' % $.dashboard.prefix,
    },
    queryFrontend:: null,
    store:: null,
    receive:: null,
    rule:: null,
    compact:: null,
    bucketReplicate:: null,
};

{
    'out/thanos_recording_rules.yaml': std.manifestYamlDoc(thanos.
prometheusRules),
    'out/thanos_alerting_rules.yaml': std.manifestYamlDoc(thanos.
prometheusAlerts),
}
+
{
    ['out/dashboard_' + name]: std.manifestJson(thanos.
grafanaDashboards[name])
    for name in std.objectFields(thanos.grafanaDashboards)
}
```

There's a lot to digest here, so let's walk through it.

First, we define a local variable named `thanos` that imports the mixin we just downloaded. Specifically, it imports the `mixin.libsonnet` file, which should be present in most mixins and handles importing all the relevant Prometheus rules and Grafana dashboards.

Next, we provide field overrides where we set each component we don't care about to null and configure the ones we do care about. The values used for `query` and `sidecar` are the same as the defaults, so they're not strictly necessary, but I included them to show how you *could* override them.

Finally, we use the syntax we learned earlier for outputting data to multiple files to create a file for recording rules, a file for alerting rules, and a file for each Grafana dashboard. Due to limitations with object comprehensions, we have to handle the dashboard as a separate object and combine it using the + operator with the object for the Prometheus rules.

Now, we can run this file through the Jsonnet command line to output all of these files. In order for Jsonnet to know where to find our mixin, though, we need to tell it to search an additional path for libraries by adding a `-J vendor/` flag. Ultimately, the command to render our configured mixin looks like this:

```
$ jsonnet -J ch11/vendor -Sm ch11/ ch11/thanos.jsonnet

ch11/out/dashboard_overview.json
ch11/out/dashboard_query.json
ch11/out/dashboard_sidecar.json
ch11/out/thanos_alerting_rules.yaml
ch11/out/thanos_recording_rules.yaml
```

We can inspect each of the five output files to confirm that they look reasonable and see all of the sweet Prometheus rules and Grafana dashboards we got – essentially – for free!

Combing through the alerting rules of Thanos Query, for instance, you'll see plenty of alerts related to things such as gRPC. This is just an example of the enormous benefit mixins provide by allowing developers to define recommended alerting rules. Take this alert as an example (actual alert modified for brevity):

```
  - "alert": "ThanosQueryGrpcServerErrorRate"
    "annotations":
      "description": "Thanos Query {{$labels.job}} is failing to
handle {{$value | humanize}}% of requests."
    "expr": |
      (
        sum by (job) (rate(grpc_server_handled_total{grpc_code=
~"Unknown|ResourceExhausted|Internal|Unavailable|DataLoss|
DeadlineExceeded", job=~".*thanos-query.*"}[5m]))
      /
```

```
        sum by (job) (rate(grpc_server_started_total{job=~".*thanos-
query.*"}[5m]))
      * 100 > 5
      )
```

You don't have to become an expert in gRPC or know the gRPC APIs that Thanos Query defines to know what you should monitor for issues. This helps to remove some of the "unknown unknowns" that are so common in running software that you didn't write yourself.

Hopefully, you've gotten to see the potential in Monitoring Mixins through this exercise and I encourage you to seek out monitoring mixins for software you're currently running to provide solid baselines of what you should be monitoring for it. I think you'll find that – similar to how there seems to be a Prometheus exporter for everything – there is an astonishing amount of monitoring mixins out there that can help ease the initial burden of ensuring that your systems are observable and you know the right signals to be looking at. Grafana's `jsonnet-libs` repository (`https://github.com/grafana/jsonnet-libs/`) is a great starting point for finding existing mixins. If a mixin doesn't exist yet for a product you use, consider creating an open sourcing one! Your work might just help make the world a little bit more reliable.

Summary

In this chapter, we've gone from zero to hero in Jsonnet. Take a deep breath and admire the fact that you've – most likely – just learned a whole new programming language. With Jsonnet, we can remove some of the tedium of managing YAML files – like those used by Prometheus – by hand. This can help to ensure more consistent and less error-prone management of these files over time by keeping our code DRY.

Not only that, but we've learned how the Monitoring Mixin specification enables projects to create reusable, configurable Prometheus rules and Grafana dashboards. By leveraging existing mixins (and/ or making our own!), we can ensure greater observability and monitoring of our systems.

In our next chapter, we'll continue exploring ways to mature our operations of Prometheus monitoring stacks using CI/CD tooling to perform tasks such as validation and testing.

Further reading

To learn more about the topics that were covered in this chapter, take a look at the following resources:

- *Using Jsonnet to Package Together Dashboards, Alerts and Exporters – Tom Wilkie (KubeCon EU 2018)*: `https://www.youtube.com/watch?v=b7-DtFfsL6E`

- *Taming the Beast: Comparing Jsonnet, Dhall, Cue*: `https://pv.wtf/posts/taming-the-beast`

- *Mixins in Object Oriented Programming*: `https://en.wikipedia.org/wiki/Mixin#History`

- *Jsonnet project documentation*: `https://jsonnet.org/`

- *Monitoring Mixins website*: `https://monitoring.mixins.dev/`

- *Monitoring Mixins design proposal*: `https://github.com/monitoring-mixins/docs/blob/5ef6256185cdc47db8139f576bf463a80d91780e/design.pdf`

Utilizing Continuous Integration (CI) Pipelines with Prometheus

As the trend toward **Infrastructure-as-Code (IaC)** continues, more and more focus is placed on ensuring that artifacts such as configuration files are version-controlled (usually via Git) and continuously synchronized. With the multiple files typically involved in a Prometheus stack – from Prometheus's configuration file to rules files to Alertmanager's configuration and everything in between – it can become difficult to ensure that every change to these files is properly vetted and safe to deploy.

Certainly, we can rely on code reviews, approvals, and branch protection rules on repositories to try to catch as many problems as we can. However, inevitably, things get missed, and systems break due to misconfigurations. The simple solution here is to remove the possibility of human error as much as possible. Through a combination of various tools, we can build out **continuous integration (CI)** pipelines to perform this validation and linting for us.

This chapter will assume that you store your Prometheus configuration and rules inside a Git repository since that is compulsory for using CI tools such as Jenkins, GitLab CI, or GitHub Actions. For this chapter, we will focus on CI integrations using GitHub Actions due to its ubiquity, convenience for local testing, and the fact that the code for this book is stored in GitHub. However, the same principles will be easily transferrable across CI tools – just with a different syntax.

With that in mind, in this chapter, we'll be covering these main topics:

- GitHub Actions
- Validation in CI
- Linting Prometheus rules with Pint

Let's get started!

Technical requirements

For this chapter, we'll be stepping away from our Prometheus deployment and doing things locally. Consequently, you'll need these tools installed to follow along:

- **docker**: `https://docs.docker.com/desktop/`
- **act**: `https://github.com/nektos/act#installation`
- **pint**: `https://github.com/cloudflare/pint/releases/tag/v0.50.0`

The code used in this chapter is available at `https://github.com/PacktPublishing/Mastering-Prometheus`.

GitHub Actions

GitHub Actions quickly carved out a place for itself in the CI ecosystem when it launched in 2018. Since GitHub is the de-facto hub for major open source projects, GitHub Actions has also seen wide adoption within those open source projects, including Prometheus. If you're not familiar with it, fear not. GitHub Action workflows are written in – you guessed it – YAML. So, at least there's not some funky Java-adjacent language that you need to learn to understand this chapter (looking at you, Jenkins).

Now, this obviously is not a book about CI tools, DevOps, or anything of the sort. So, why are we still talking about GitHub Actions? Well, if you're already familiar with it, then skip right ahead to the next section. However, if you're not, let's go over some basics so that we're all using the same terminology and the rest of the chapter will make sense to you.

GitHub Actions is just the name of the **Continuous Integration/Continuous Delivery** (**CI/CD**) platform provided by GitHub. The thing that GitHub Actions runs is not called an "action" (although some may refer to it as such). Instead, GitHub Actions runs **workflows**, which are suites of one or more **jobs** that are run in response to some event. The "event" tends to be either a pull request to a Git repository or a push to one.

Each job consists of multiple **steps** that are sequentially executed. Each step is a command (as in, a shell command) or a call to an **action**. In this sense, an action is a custom application specific to GitHub Actions and handles running a complex – but repeatable – task.

When multiple jobs are defined in a workflow, they will run in parallel by default. However, you can explicitly specify dependencies so that a job will not start until one it depends on completes. For example, with Prometheus, you may have a job that replaces sensitive values in your configuration file and a job that validates it. The validation job can be set to depend on the substitution job so that it won't run if the substitution fails. Of course, you could always make those two things sequential steps within the same job. It typically boils down to personal preference.

Finally, each job is executed by a **runner**. A runner is simply a server that executes jobs. They can be self-hosted, or you can use the ones provided by GitHub (for a fee, of course). However, for our purposes, we'll be using a tool called `act` to run our workflows locally.

The `act` tool works by emulating a GitHub Actions runner via a Docker container. It handles the injection and processing of a variety of special environment variables and other things available within a normal GitHub Actions runner, which enables you to test and run most CI workflows locally. With that installed, let's build some workflows to see what we can do for Prometheus in CI!

Validation in CI

A variety of tools exist for testing and validating various aspects of Prometheus's configuration files. However, the two most simple and direct options are to use the `promtool` and `amtool` binaries for performing validations related to Prometheus and Alertmanager, respectively. The Prometheus project provides these binaries, which are automatically included in releases downloaded from GitHub for their respective projects.

Each of these tools has multiple subcommands that can be used to test and validate different parts of their related configurations. To begin with, let's see what we can accomplish using the `promtool` CLI application.

Using promtool

The `promtool` binary is extremely flexible in a number of things it can do. It's able to do everything from interacting directly with the TSDB to running `pprof` profiles. However, we're going to focus on what it can do when it comes to validating Prometheus's configuration and rules files.

For testing the actual Prometheus configuration file, `promtool` has the `check config` subcommand. For validating rules, it has the `check rules` subcommand. Finally, for running unit tests on rules (see *Chapter 5*), it has the `test rules` subcommand.

Let's see what it would look like to run all three of these in a GitHub Actions workflow. First, we'll start by validating the Prometheus configuration file. We'll validate a basic `prometheus.yaml` file with these contents:

```
global:
  scrape_interval: 30s
  scrape_timeout: 10s
  evaluation_interval: 30s
alerting:
  alertmanagers: []
rule_files: []
scrape_configs:
- job_name: prometheus
```

```
  static_configs:
    - targets:
        - "localhost:9090"
      labels:
        service: "prometheus"
```

We'll run our workflow inside of a Docker image for Prometheus, so the `promtool` binary will already be available to us. All we need to do is add a step to run `promtool check config` on this file. The workflow file (`promtool.yaml`) looks like this:

```
name: promtool
on: push

jobs:
  config:
    name: Validate Prometheus Config
    runs-on: ubuntu-latest
    container: quay.io/prometheus/prometheus:v2.46.0
    steps:
      - name: Checkout/clone code
        uses: actions/checkout@v4

      - name: Validate Prometheus config file
        run: promtool check config ch12/prometheus.yaml
```

Finally, we can run this using `act` to confirm that our configuration file is valid.

> **Note**
>
> If this is your first time running `act`, you will be prompted for the Docker image you want to use as the default for the GitHub Action runners it simulates. You can choose the `medium` image.
>
> Additionally, `act` assumes that you are running it from the root of the relevant Git repository. To use a different directory, use the `-C` flag.

To run `act`, we'll provide the path to the workflow, and it will handle the rest:

```
$ act -W ch12/workflows/promtool.yaml
```

The last few lines of output should look like this:

```
[promtool/Validate Prometheus Config]  ☆ Run Main Validate Prometheus
config file
[promtool/Validate Prometheus Config]     docker exec cmd=[sh -e /
var/run/act/workflow/1.sh] user= workdir=
| Checking ch12/prometheus.yaml
```

```
|  SUCCESS: ch12/prometheus.yaml is valid prometheus config file
syntax
|
[promtool/Validate Prometheus Config]    ✅  Success - Main Validate
Prometheus config file
[promtool/Validate Prometheus Config] Cleaning up container for job
Validate Prometheus Config
[promtool/Validate Prometheus Config] 🏴  Job succeeded
```

Huzzah! Our configuration is valid. Now, what about rules? We'll copy the rules file we used in *Chapter 5* (test-rule.yaml), which looks like this:

```
groups:
  - name: chapter5
    rules:
      - alert: SSHDown
        expr: >
          max_over_time(
            probe_success{job="blackbox_ssh"}[5m]
          ) != 1
        for: 5m
        labels:
          severity: warning
          team: sre
        annotations:
          description: "Cannot SSH to {{ $labels.instance }}"
          grafana: https://grafana.example.com/ssh-dashboard?var-
instance={{ $labels.instance }}
```

If you're using the code for this chapter from the GitHub repository for this book, I'll include a rules file containing the alerting rules that are in the Kubernetes-based Prometheus environment we've been using throughout the book. However, since that's over 750 lines of YAML, it is omitted here.

Now, to validate rules, we'll add another job to our workflow. We'll use a separate job since the validation of the config and the validation of rules files are not dependent on each other. To add the new job, append this to your promtool.yaml file:

```
rules:
  name: Validate Prometheus Rules
  runs-on: ubuntu-latest
  container: quay.io/prometheus/prometheus:v2.46.0
  steps:
    - name: Checkout/clone code
      uses: actions/checkout@v4
```

```
        - name: Validate Prometheus rules files
          run: promtool check rules ch12/rules/*.yaml
```

Running with `act` again, we'll still see output from the `config` job sprinkled in since the two jobs run in parallel, but we should also see lines of output that look something like this:

```
[promtool/Validate Prometheus Rules]    ✅    Success - Main Validate
Prometheus config file
[promtool/Validate Prometheus Rules]  ☆ Run Main Validate Prometheus
rules files
[promtool/Validate Prometheus Rules]    🐳    docker exec cmd=[sh -e /
var/run/act/workflow/2.sh] user= workdir=
| Checking ch12/rules/rules.yaml
|   SUCCESS: 57 rules found
|
| Checking ch12/rules/test-rule.yaml
|   SUCCESS: 1 rules found
|
[promtool/Validate Prometheus Rules]    ✅    Success - Main Validate
Prometheus rules files
[promtool/Validate Prometheus Rules] Cleaning up container for job
Validate Prometheus Rules
[promtool/Validate Prometheus Rules]  🏁   Job succeeded
```

We're getting the hang of this! Now, all that's left is to ensure we run any unit tests on our rules, too. We'll slightly modify and reuse the unit test for `test-rule.yaml` from *Chapter 5* as well. Put the test in a new file called `unit-test.yaml` with these contents:

```
rule_files:
  - ../rules/test-rule.yaml

evaluation_interval: 1m
tests:
  - interval: 1m
    input_series:
      - series: 'probe_success{instance="server1", job="blackbox_
ssh"}'
        values: "1 1 1 1 0+0x12"
      - series: 'probe_success{instance="server2", job="blackbox_
ssh"}'
        values: "1 1 1 _ 1+0x12"
    alert_rule_test:
      - alertname: SSHDown
        eval_time: 9m
        exp_alerts: []
      - alertname: SSHDown
```

```
        eval_time: 14m
        exp_alerts:
          - exp_labels:
              severity: warning
              team: sre
              instance: server1
              job: blackbox_ssh
            exp_annotations:
              description: "Cannot SSH to server1"
              grafana: "https://grafana.example.com/ssh-dashboard?var-
  instance=server1"
```

Lastly, we'll add one more step to the `rules` job in our `promtool.yaml` workflow file to run `promtool test rules`:

```
      - name: Run unit tests on Prometheus rules files
        run: promtool test rules ch12/tests/*.yaml
```

Running this workflow one last time with `act`, the final few lines of output now show the unit test succeeding in addition to all of our prior success:

```
[promtool/Validate Prometheus Rules]      ✓   Success - Main Validate
Prometheus rules files
[promtool/Validate Prometheus Rules]  ☆ Run Main Run unit tests on
Prometheus rules files
[promtool/Validate Prometheus Rules]      🐳   docker exec cmd=[sh -e /
var/run/act/workflow/3.sh] user= workdir=
| Unit Testing:  ch12/tests/unit-test.yaml
|   SUCCESS
|
[promtool/Validate Prometheus Rules]      ✓   Success - Main Run unit
tests on Prometheus rules files
[promtool/Validate Prometheus Rules] Cleaning up container for job
Validate Prometheus Rules
[promtool/Validate Prometheus Rules] 🏁   Job succeeded
```

Those three steps in your CI pipeline will help to catch a plethora of potential errors in your Prometheus files. They are a great starting point, and I recommend them to anyone, even if you're the only one making changes to the files! They've saved me from breaking things plenty of times.

Although this provides a good amount of base coverage for Prometheus, we still need to apply the same level of validation to Alertmanager. Next up, we'll explore how to build out a workflow for Alertmanager using `amtool`.

Using amtool

Our workflow for Alertmanager's CI pipeline is going to look largely similar in structure to our Prometheus pipeline. We'll still just run one job with multiple steps. The steps in this case are `amtool check-config` and `amtool config routes test`.

The `check-config` command does exactly what it sounds like, and is akin to the `promtool check config` command in Prometheus. Why is the subcommand syntax different, you may ask? I don't know – it bothers me, too.

To get our workflow started, we'll need to create the configuration that we'll be validating. In a new file named `alertmanager.yaml`, add these contents:

```yaml
route:
  receiver: "fallthrough"
  routes:
    - receiver: "slack-devops"
      matchers:
        - "team = devops"
      routes:
        - receiver: "pagerduty-devops"
          matchers:
            - "environment = production"
    - receiver: "sre"
      matchers:
        - "environment = production"
        - "team = sre"
receivers:
  - name: "sre"
    slack_configs:
      - channel: "#sre-alerts"
        api_url: "https://hooks.slack.com/services/ZZZZZZZZZZZZZZZ"
  - name: "slack-devops"
    slack_configs:
      - channel: "#devops-alerts"
        api_url: "https://hooks.slack.com/services/ZZZZZZZZZZZZZZZ"
  - name: "pagerduty-devops"
    pagerduty_configs:
      - routing_key: "XXXXXXXXXXXXXXX"
  - name: 'fallthrough'
    webhook_configs:
      - url: 'http://127.0.0.1:5001/'
```

Now, in a new file called `amtool.yaml`, we'll build out the workflow. To get started, we'll just validate the configuration file. The workflow file looks like this:

```
name: amtool
on: push

jobs:
  config:
    name: Validate Alertmanager config
    runs-on: ubuntu-latest
    container: quay.io/prometheus/alertmanager:v0.25.0
    steps:
      - name: Checkout/clone code
        uses: actions/checkout@v4

      - name: Validate Alertmanager config file
        run: amtool check-config ch12/alertmanager.yaml
```

Next, we'll run this new workflow through `act`, like so:

```
$ act -W ch12/workflows/amtool.yaml
```

And our output should look something like this:

```
[amtool/Validate Alertmanager config]  ☆ Run Main Validate
Alertmanager config file
[amtool/Validate Alertmanager config]     🐳   docker exec cmd=[sh -e /
var/run/act/workflow/1.sh] user= workdir=
| Checking 'ch12/alertmanager.yaml'  SUCCESS
| Found:
|  - global config
|  - route
|  - 0 inhibit rules
|  - 4 receivers
|  - 0 templates
|
[amtool/Validate Alertmanager config]     ✅   Success - Main Validate
Alertmanager config file
[amtool/Validate Alertmanager config] Cleaning up container for job
Validate Alertmanager config
[amtool/Validate Alertmanager config] 🏳  Job succeeded
```

Hooray! I wrote a valid config. That's all well and good, but what if we could do more than just validate the config?

One of the most common issues I see with people managing Alertmanager configs is difficulty grasping how the routing tree works within the config. A natural inclination is to just add a new route to the end of the configuration file, but that doesn't always work how people expect since an earlier route may catch the alert before it gets that far. To prevent introducing regressions and confirm desired behavior, we can utilize a lesser-known feature of `amtool` to test our routes based on label matchers.

To simplify things, we'll just test one route directly in the workflow. In practice, you may want to create a simple shell script to do this validation for a large number of routes and label combinations. In our `amtool.yaml` workflow file, we'll append a new job that looks like this:

```
routes:
  name: Validate Alertmanager routes
  runs-on: ubuntu-latest
  container: quay.io/prometheus/alertmanager:v0.25.0
  steps:
    - name: Checkout/clone code
      uses: actions/checkout@v4

    - name: Validate Alertmanager Route
      run: |
        amtool config routes \
          --config.file=ch12/alertmanager.yaml \
          test --verify.receivers \
          pagerduty-devops \
          team=devops environment=production \
          service=jenkins
```

Finally, running this with `act` will give us output that looks like this:

```
[amtool/Validate Alertmanager routes] ☆ Run Main Validate
Alertmanager Route
[amtool/Validate Alertmanager routes]     🐳  docker exec cmd=[sh -e /
var/run/act/workflow/1.sh] user= workdir=
| pagerduty-devops
[amtool/Validate Alertmanager routes]    ✅   Success - Main Validate
Alertmanager Route
[amtool/Validate Alertmanager routes] Cleaning up container for job
Validate Alertmanager routes
[amtool/Validate Alertmanager routes] 🏁   Job succeeded
```

Hopefully, you have a good idea by now on how to implement some basic safety checks in your CI pipelines in order to ensure that your Prometheus and Alertmanager configurations are valid and working as expected. But there's one more tool that I want to cover before we wrap up that can really take this validation to the next level, so let's talk about Pint next!

Linting Prometheus rules with Pint

Pint is the secret sauce we've been missing all along. Honestly, if you add linting via the `pint` command to your CI pipelines, you can probably just skip most of the `promtool` validation for rules. But what is it? A unit of liquid measurement? Yes, and more!

Pint is a delightful tool to come out of Cloudflare, where they run Prometheus deployments at a tremendous scale (see the blog posts linked at the end of the chapter). In order to accommodate this scale, the team responsible for managing these hundreds of Prometheus instances needs to avoid being a bottleneck for other teams needing to manage their recording and alerting rules on these servers. In order to ensure that their rules are safe, valid, and will work as intended, they created Pint.

The core feature of Pint that distinguishes it from something such as `promtool` is that it connects to one or more active Prometheus instances to confirm that rules will work as expected. It does this by evaluating the PromQL expressions within rule definitions to catch issues such as poor label matchers, non-existent time series, and queries with vector matching (see *Chapter 3*), where the labels differ between series on the left and right sides.

In addition to the normal `promtool` validation of ensuring that YAML is formatted correctly and the expected keys are present in all the rules, Pint adds checks for the following:

- Enforcing specific annotations on alerting rules
- Enforcing a comparison operator (for example, <, >, and so on) in alerting rules
- Warning if a new/updated alerting rule would've fired a bunch of alerts recently
- Enforcing that any external labels from the Prometheus server used in templates actually exist on the Prometheus server
- Validating templates in labels and annotations on alerting rules
- Conflicting labels between Prometheus' external labels and those added to an alert
- Enforcing that aggregation functions such as `sum`, `count`, and so on must preserve certain labels
- Warning if queries are using metrics that don't exist
- Validating that links in annotations actually work
- Warning if range queries are requesting a time range greater than the amount of time that Prometheus is configured to store data

And that's only some of the checks! When used properly, Pint unlocks a brand new suite of integration tests for your Prometheus rules, allowing you to confidently deploy your alerts. Let's see how we can make that dream a reality.

Configuring Pint

Pint is configured using **HashiCorp Configuration Language** (**HCL**), which some of you may be familiar with through open source projects such as Terraform or Packer. By default, Pint uses a configuration file named `.pint.hcl` in whichever directory the command is run. Consequently, that's the file we'll create and begin defining.

> **Note**
>
> Configuring and running Pint is expected to be done locally for this part of the chapter. Due to the fact that our Prometheus environment is not externally accessible and, therefore, cannot be used by GitHub Actions, Pint would be unable to run queries against it, and its usefulness would be diminished.

The first thing we need to add to our Pint configuration file is a block for configuring the Prometheus instances that will be used to aid in linting our rules. For our purposes, we'll only configure one Prometheus instance. In production environments, you can (and should) configure multiple Prometheus instances and use the `include` or `exclude` keywords within their definition to specify a list of rules files that should be included or excluded for testing against the given Prometheus instances.

To configure our Prometheus instance, we only need to specify the URL to access it. Since we will port forward locally to our Prometheus service in the Kubernetes environment we've used thus far, we can use `localhost` as the URL. It looks like this:

```
prometheus "mastering-prometheus" {
  uri          = "http://localhost:9090"
}
```

Now, if we use our test rule from earlier and run `pint lint test-rule.yaml`, we should see this:

```
$ pint lint rules/test-rule.yaml

level=INFO msg="Loading configuration file" path=.pint.hcl
level=INFO msg="Finding all rules to check" paths=["rules/test-rule.
yaml"]
level=INFO msg="Configured new Prometheus server" name=mastering-
prometheus uris=1 uptime=up tags=[] include=[] exclude=[]
rules/test-rule.yaml:5-8 Bug: `mastering-prometheus` Prometheus server
at http://localhost:9090 didn't have any series for `probe_success`
metric in the last 1w. (promql/series)
5 |          expr: >
6 |            max_over_time(
7 |              probe_success{job="blackbox_ssh"}[5m]
8 |            ) != 1
```

```
level=INFO msg="Problems found" Bug=1
level=ERROR msg="Execution completed with error(s)" err="found 1
problem(s) with severity Bug or higher"
```

Whereas `promtool` reported earlier that this rule was fine, Pint caught on to the fact that there is no time series for the metric we're querying! That's just with the basic suite of rules, too. Let's see what we can do by defining our own custom rules.

Disabling checks

We expect there to be no `probe_success` time series since we're just using it as a test and, therefore, don't need to be reminded of it every time we run Pint. Consequently, we can disable linting for that specific check on that particular rule by adding a comment that will cause Pint to skip that check. Try adding `# pint disable promql/series(probe_success)` just before the list entry for that alert and rerun Pint to confirm that no errors are displayed.

To configure additional rules, we can add repeated `rule` block definitions. A common pattern for Prometheus alerts is to have a label for a `severity` level with a value of `critical`, `warning`, or `info`. We can enforce that using Pint via a `rule` block definition like this:

```
rule {
  match {
    kind = "alerting"
  }
  label "severity" {
    severity = "bug"
    value    = "(critical|warning|info)"
    required = true
  }
}
```

In this configuration, we tell Pint to match all alerting rules and enforce the rules within the block on them. The preceding example enforces that rules must have a `severity` label, and it must match the regex within the `value` field.

Pint regex behavior

As with Prometheus, Pint automatically anchors regexes. This means that the `^` symbol is prepended, and the `$` symbol is appended to the regex so that the regex must match the whole line from start to finish. If there are multiple acceptable patterns for a field, you can use `values` instead of `value` to specify a **list** of acceptable regex patterns that the value can match. If at least one successfully matches, the check will pass.

Each Pint `rule` block can have multiple rules contained within it, which are all enforced against the matched Prometheus rules. Expanding on the preceding example, let's define some required annotations as well.

The *Monitoring Mixins* spec (see *Chapter 11*) defines all alerts as requiring a severity label with a value of `critical`, `warning`, or `info`, but also requires `summary` and `description` annotations. A `summary` annotation should be a single sentence (starts with a capital letter and ends with a period) that is a slightly more detailed version of the alert name. On the other hand, a `description` annotation is a fully detailed description of the alert that gives more of the specifics of the alert, including templated values for information such as the server (`{{ $labels.instance }}`) the alert applies to or the specific value (`{{ $value }}`) returned by evaluating the PromQL expression.

Adding to our existing `rule` block, we can add the following to enforce both of these:

```
annotation "summary" {
  severity = "bug"
  required = true
  value = "[A-Z][^{.]+.\\."
}

annotation "description" {
  severity = "bug"
  required = true
  value = "[A-Z].+\\{\\{.*\\}\\}.*\\."
}
```

For the `summary` annotation rule, we enforce that the summary must begin with a capital letter, end with a period, be a single sentence (cannot contain any periods prior to the end, exclamation points, or question marks), and cannot contain any templated values. We skip the single-sentence requirement for the `description` annotation rule but still enforce that it begins with a capital letter and ends in a period. Additionally, we enforce that it *must* have at least one templated value within it – otherwise, what's the benefit of a `description` annotation compared to a `summary` annotation?

Rerunning Pint, we'll see new errors pop up:

```
$ pint lint rules/test-rule.yaml

level=INFO msg="Loading configuration file" path=.pint.hcl
level=INFO msg="Finding all rules to check" paths=["rules/test-rule.
yaml"]
level=INFO msg="Configured new Prometheus server" name=mastering-
prometheus uris=1 uptime=up tags=[] include=[] exclude=[]
rules/test-rule.yaml:14-16 Bug: `summary` annotation is required.
(alerts/annotation)
 14 |         annotations:
 15 |            description: "Cannot SSH to {{ $labels.instance }}"
```

```
 16 |             grafana: https://grafana.example.com/ssh-
dashboard?var-instance={{ $labels.instance }}
```

```
rules/test-rule.yaml:15 Bug: `description` annotation value must match
`^[A-Z].+\{\{.*\}\}.*\.$`. (alerts/annotation)
 15 |             description: "Cannot SSH to {{ $labels.instance }}"
```

```
level=INFO msg="Problems found" Bug=2
level=ERROR msg="Execution completed with error(s)" err="found 1
problem(s) with severity Bug or higher"
```

Pint caught that our test rule does not have a `summary` annotation and that our `description` annotation doesn't end in a period. To fix those, you can update the test rule definition (I'll put mine in a new `test-rule-updated.yaml` file) so that the `annotations` block looks like this:

```
    annotations:
        summary: "No successful SSH probes in the last 5 minutes."
        description: "Cannot SSH to {{ $labels.instance }}."
        grafana: https://grafana.example.com/ssh-dashboard?var-
instance={{ $labels.instance }}
```

Running Pint one final time, we should see that all checks now pass against our test rule:

```
$ pint lint rules/test-rule-updated.yaml
```

```
level=INFO msg="Loading configuration file" path=.pint.hcl
level=INFO msg="Finding all rules to check" paths=["rules/test-rule-
updated.yaml"]
level=INFO msg="Configured new Prometheus server" name=mastering-
prometheus uris=1 uptime=up tags=[] include=[] exclude=[]
```

There are a variety of different, customizable checks that you can run with Pint to perform these same sorts of dynamic evaluations. I recommend exploring the Pint documentation linked at the end of the chapter for more examples of how to accomplish things such as automatically checking URLs in annotations to confirm they work (the `rule/link` check) or raising warnings about new alerts that have fired too often recently (the `alerts/count` check).

Integrating Pint with CI

Given that this is a chapter on CI, I'd be remiss if I didn't at least touch on how Pint can be integrated into your CI workflows, even more so because Pint is *designed* to be run in CI and includes valuable features only available within CI.

When running Pint in a CI environment, in addition to the normal output we've seen thus far, it is able to add *comments* to pull requests on the specific lines that are raising errors. Additionally, it is able to automatically run checks against only the files changed in a given pull request, which helps to prevent users from seeing errors unrelated to their actual changes.

To configure Pint for CI, we configure a `ci` block in the same `.pint.hcl` file we've used thus far. All that is needed is to tell Pint which files to include in CI checks, and (optionally) the base branch for the repository (for example, `main` or `master`). It looks like this:

```
ci {
  include = [ ".github/pint/test-rule.yaml" ]
  baseBranch = "main"
}
```

Optionally, you may also configure a `repository` block. However, that is generally unnecessary since its configuration values will automatically be derived from environment variables set by GitHub Actions runners.

> **Restricting files to check**
>
> Although the `ci` block enables us to configure which files Pint will check, it is important to note that it only applies to **pull requests**. If you enable Pint to run on *pushes* to a branch, Pint will run against all files it can discover. To ensure that only expected files are checked, set the `working-directory` value on your GitHub Actions workflow job step to the subdirectory containing your rules, or skip running Pint on pushes altogether.

Here's an example GitHub workflow used by this book:

```
name: pint
on:
  pull_request:
    branches:
      - "main"
    paths:
      - ".github/pint/test-rule.yaml"


jobs:
  pint:
    name: Pint CI test
    runs-on: ubuntu-latest
    steps:
      - name: Checkout/clone code
        uses: actions/checkout@v4

      - name: Run pint
```

```
uses: prymitive/pint-action@v1
with:
  token: ${{ github.token }}
  config: '.github/pint/pint.hcl'
```

Notably, this workflow uses a reusable GitHub Action specific to Pint created by its author, which simplifies the usage of Pint within this workflow. When run against our `test-rule.yaml` file from earlier, we can observe comments added by Pint to the GitHub pull request:

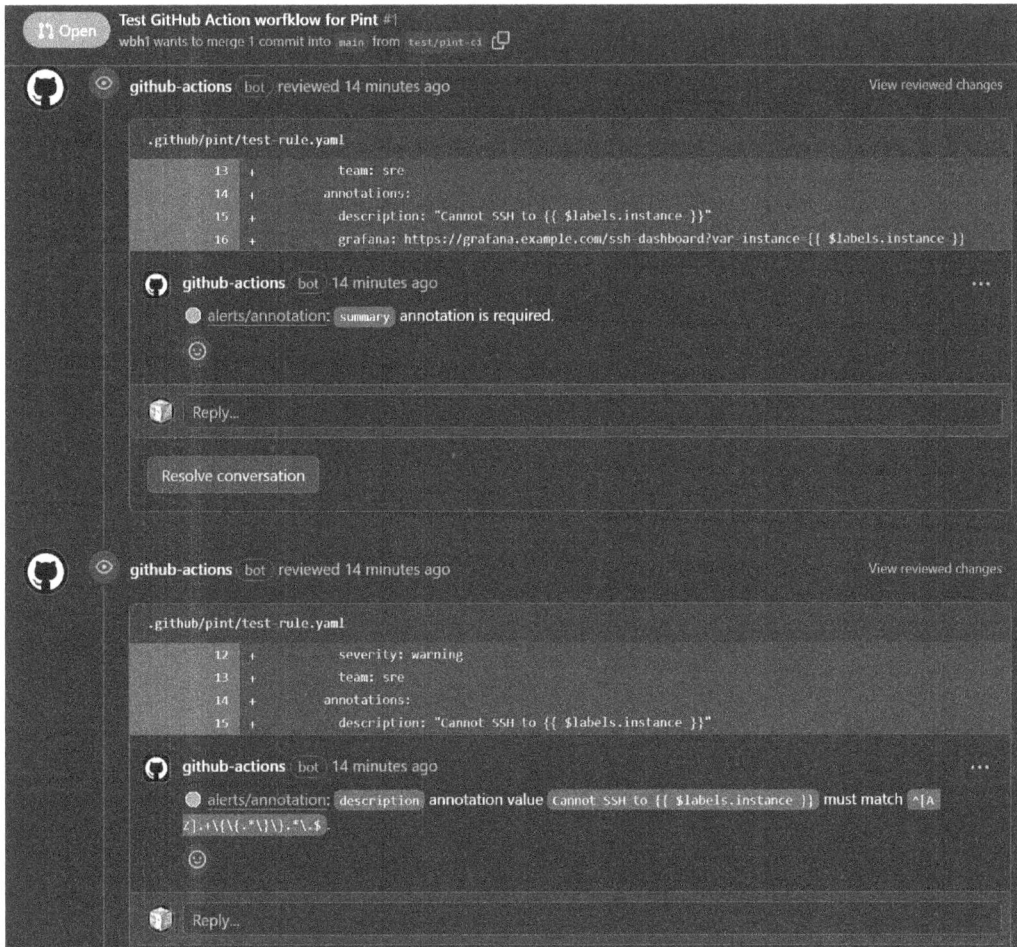

Figure 12.1 – Pint CI comments on a GitHub pull request

To fully see what Pint output looks like when run against a pull request via CI, check out the example on this book's repository at `https://github.com/PacktPublishing/Mastering-Prometheus/pull/1`.

Running Pint in CI like this is the final step toward automatically and comprehensively validating your Prometheus rules. The specific checks you configure for Pint will vary depending on your environment's needs. I once again encourage you to consult the Pint documentation linked at the end of this chapter to discover additional ways you may be able to leverage Pint in your environment. Nevertheless, even with the default test suite, Pint provides enormous value over solely using `promtool`.

Summary

In this chapter, we've explored how to ensure that our Prometheus configuration, Prometheus rules, and Alertmanager configuration are properly validated using CI pipelines with GitHub Actions. This is a critical step toward the maturity of your Prometheus environment and enabling other teams who may not be Prometheus **Subject Matter Experts** (**SMEs**) to contribute to the monitoring environment without fear.

CI environments vary widely across the industry as there are a plethora of different CI tools available for use. But whether you use GitHub Actions, Jenkins, CircleCI, GitLab, or anything else, I hope this chapter has given you ideas on how to better integrate validation steps for your Prometheus environment into your CI tool of choice.

In our next chapter, we'll continue exploring ways to build out mature monitoring environments by looking at how we can define and implement **Service Level Objectives** (**SLOs**) using Prometheus.

Further reading

To learn more about the topics that were covered in this chapter, take a look at the following resources:

- *GitHub Actions workflow syntax*: `https://docs.github.com/en/actions/using-workflows/workflow-syntax-for-github-actions`
- *Cloudflare Prometheus blog posts*:
  - `https://blog.cloudflare.com/how-cloudflare-runs-prometheus-at-scale`
  - `https://blog.cloudflare.com/monitoring-our-monitoring/`
- *Pint documentation*: `https://cloudflare.github.io/pint/configuration.html`
- *Monitoring Mixins labels and annotations*: `https://github.com/monitoring-mixins/docs#guidelines-for-alert-names-labels-and-annotations`

13

Defining and Alerting on SLOs

Few topics in the observability space are currently as exciting as the increased focus on **service-level objectives** (SLOs). The interest around SLOs has reached such a fever pitch that there is even an annual conference – SLOConf, started in 2021 – entirely dedicated to them. SLOs are intrinsically intertwined with observability technologies such as Prometheus, and learning how to utilize Prometheus data in your SLOs will help you take your Prometheus environment to a new level.

SLOs are a quintessential example of how observability and its associated telemetry provide direct business value – not just operational support. At their core, SLOs provide a consistent, measurable way to see how your services are performing at a high level. This simple, numeric measurement is ideal for executives and business leaders to be able to report on the health and reliability of the services they are ultimately responsible for. Since Prometheus is a metrics system, which is all about subjecting things to numeric measurements, it is an ideal fit for an SLO data source.

With that in mind, in this chapter, we'll be covering these main topics:

- Understanding SLIs, SLOs, and SLAs
- Defining SLOs with Prometheus data
- Using Sloth and Pyrra for SLOs

Let's get started!

Technical requirements

For this chapter, we'll be leveraging our Prometheus environment that was built in *Chapter 2*. In addition to that, you'll need these tools installed to follow along:

- **sloth**: `https://github.com/slok/sloth/releases/tag/v0.11.0`
- **pyrra**: `https://github.com/pyrra-dev/pyrra/releases/tag/v0.7.2`

The code used in this chapter is available at `https://github.com/PacktPublishing/Mastering-Prometheus`.

Understanding SLIs, SLOs, and SLAs

Service-level objectives are part of a suite of other service-level acronyms, including **service-level indicators** (**SLIs**) and **service-level agreements** (**SLAs**). To try to explain it simply, SLIs are the lowest level of the three. They consist of individual, raw metrics (hint: this is where Prometheus will be helpful). For example, a count of error responses from your application. SLOs are comprised of one or more SLIs that, when taken together, form a meaningful measurement of performance measured on a scale of 0–100%. Generally, they are measured in "nines," representing how many number 9s are in the percentage – 99.9% is three nines, 99.99% is four nines, and so on. An example SLO could be an objective that 99.9% of the requests to your application will be completed without an error. Finally, SLAs are SLOs with a contractual agreement thrown in. Generally, those contracts involve penalties (e.g., service credits) for failure to meet the specified level of performance:

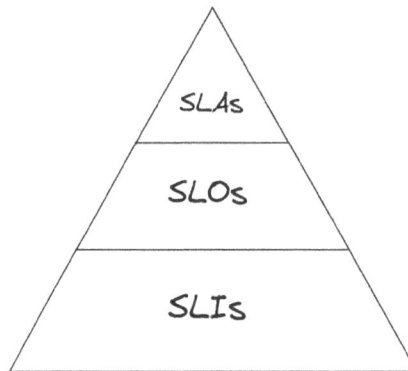

Figure 13.1 – Service level pyramid

Primarily, we will focus on SLOs in this chapter. Inherently, that will mean that we will be working with SLIs as well. However, we won't be discussing SLAs much more. That is not to imply that SLAs are not important, but they are more of a business-focused topic, whereas this book is more technical in nature. In any case, as we can see in the preceding pyramid, SLAs build off of SLOs, so everything we discuss about SLOs can still be useful in the process of defining and measuring SLAs.

> **An acronym by any other name would smell as sweet!**
>
> You may also have heard the terms **objectives and key results** (**OKRs**) or **key performance indicators** (**KPIs**). These are similar in concept to SLOs and SLIs but are generally targeted at a different audience. KPIs typically map well to SLIs, whereas OKRs are more akin to SLOs.

You may be wondering how this ties into a Prometheus book, though. The data for SLIs have to come from somewhere, and Prometheus can be that "somewhere!" SLIs are any metric data point that can be used as an indicator of how that service is performing. That doesn't mean that any Prometheus metric makes a good SLI, though. Consider the informational-style metrics we've discussed in the past, such as `prometheus_build_info`. That's not a good SLI since it tells nothing about how our service is performing. However, a metric such as `prometheus_http_requests_total` makes for a great SLI since it can tell us how many requests Prometheus is receiving and how it's responding to them.

Why SLOs matter

SLOs are integral to work in fields such as **site reliability engineering** (**SRE**). So much so that they are listed as the first "foundation" of SRE in Google's "*Site Reliability Workbook*," and three of the first five chapters of that book are dedicated to SLOs. Still, with the talk of how SLOs relate to traditional "business-y" terms, such as OKRs and KPIs, it may be tempting to dismiss them outright as yet another futile attempt by out-of-touch management to add numbers to things just for the sake of adding more numbers to graph and pester you about. However, in my experience, SLOs tend to be most effective when they are defined from the bottom up instead of the top down. In other words, the SLOs for a service should be defined by the team operationally responsible for the service – not by the director or executive that the team reports to.

That is not to say that directors and above should not have input or signoff on the SLOs their teams set. On the contrary, SLOs should be a collaborative effort bridging the gap between technical teams and the business. This is necessary in order to fully adhere to the idea that SLOs should be strict. Exceeding your **error budget** on an SLO should necessitate immediate action to restore and improve the reliability of the affected service.

> **Error budgeting**
>
> What's an error budget? Simply put, it's the amount of errors or downtime that is permissible within a given time window while still meeting your SLO. How quickly you are consuming your error budget within a given time window is referred to as your burn rate. We'll cover both of these terms more when we get to our section dealing with alerting on SLOs.

The remediation of breached SLOs typically entails forgoing other projects or feature work in order to focus on reliability efforts. In most organizations, the flexibility to drastically shift priorities on the fly like that will require buy-in and signoff from leadership. Consequently, that same leadership needs to be informed of and approve the SLOs that a team sets so that there is formal agreement on the expected availability of the services that a team owns.

SLOs are, first and foremost, a way for teams to measure themselves and their services. There is no penalty for failing to meet an SLO the way there is with an SLA. However, they can still function as the measure of excellence that a team endeavors to achieve. SLOs are not idealistic, though. Everyone says they want 100% uptime, but that's never a reasonable or achievable goal. There are simply too many outside factors that are uncontrollable, such as network or power outages, hardware failures, or dependencies on other downstream services. So, set your SLOs realistically, not aspirationally.

Additionally, SLOs need not be static; rather, they should be reviewed regularly and adjusted as needed. In cases where the SLO is being consistently exceeded, it's not uncommon to hear of teams (e.g., at Google) purposely causing controlled service outages to ensure that the consumers of their service do not end up relying on it to have a higher uptime than the team itself aspires to achieve.

The purpose of this chapter is not to convince you of the need for SLOs, though. Plenty of other books, blog posts, conference talks, and podcasts already exist for that (see the *Further reading* section at the end of this chapter). So, let's continue building the base knowledge we'll need to implement SLOs using Prometheus data.

Types of SLOs

Before we can define SLOs with Prometheus data, we'll first need to decide what **type** of SLO we'll be defining. SLOs are typically divided into two categories: **request-based** SLOs and **window-based** SLOs.

Request-based SLOs are by far the most common type of SLO. They are also the simplest. Request-based SLOs are represented as a ratio of "good" requests to "total" requests. An easy way to accomplish this is to subtract the portion of "bad" requests from 1 and multiply by 100 to get a percentage (from 0–100%) of how many good requests there were.

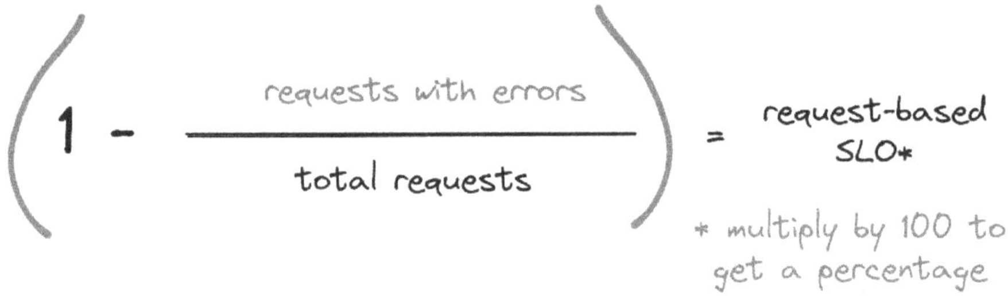

$$\left(1 - \frac{\text{requests with errors}}{\text{total requests}} \right) = \begin{array}{c} \text{request-based} \\ \text{SLO*} \end{array}$$

* multiply by 100 to get a percentage

Figure 13.2 – Request-based SLO formula

Request-based SLOs like this are an excellent fit for most services that receive a steady stream of traffic. This allows each request to be weighted similarly to other subsequent requests, and you don't have to worry about anomalous traffic artificially eating through your SLO's error budget.

In Prometheus, a robust query for a request-based SLO may look like the following:

```
clamp_min(
    sum without (pod, instance) (increase(prometheus_http_request_
duration_seconds_bucket{le="0.1"}[5m])),
    1
)
```

```
/ ignoring (le)
  clamp_min(
     sum without (pod, instance) (increase(prometheus_http_request_
duration_seconds_bucket{le="+Inf"}[5m])),
     1
  )
```

In this query, we look at the proportion of HTTP requests handled by Prometheus in a 5-minute range that were serviced in <=100ms versus the total number of HTTP requests handled during the same range. This provides us with a value between 0.0 and 1.0 that represents what percentage of requests were handled that quickly by each individual HTTP endpoint. To make it more robust, the clamp_min function is used to ensure that every endpoint receives a value, even if no requests were received on that endpoint in the given time window. In other words, it allows us to report that 100% of requests on an endpoint were serviced within the threshold, even if no requests were serviced during that time frame.

Hold your emails

There is an edge case in this query when there is only one request on an endpoint in a given time frame. In that situation, the SLO would measure at 100% even if the one request was slower than 100ms due to the clamp_min function on the top part of the query. This can be worked around by adding additional PromQL, but it is omitted for simplicity. If you're interested in the PromQL query that handles the edge case, consult this chapter's folder in the GitHub repository for this book.

Low-traffic services (think of ones where you may be measuring requests per minute rather than per second) tend to have issues with request-based SLOs. If a low-traffic service has an SLO of 99.9% availability but only receives 10 requests in a given window, then just one failed request would drop the availability measurement to 90%. For these types of services, window-based SLOs tend to be a better fit.

Window-based SLOs measure performance by time windows instead of by requests. Instead of comparing good requests to bad requests, you set criteria to distinguish a good time period from a bad time period. Consequently, you end up with two targets/thresholds: a target to determine if a time window is good or bad and an SLO target to state your desired ratio of good windows to bad windows.

Generally, these time periods are still relatively short – maybe 15 minutes at the higher end – but you need to account for the decreased data resolution when defining your SLO. Even if you're scraping Prometheus metrics every 30 seconds, if your window-based SLO is using 15-minute windows, then you really only have a data point (whether the window was good or bad) every 15 minutes. You can see how this plays out in the following diagram:

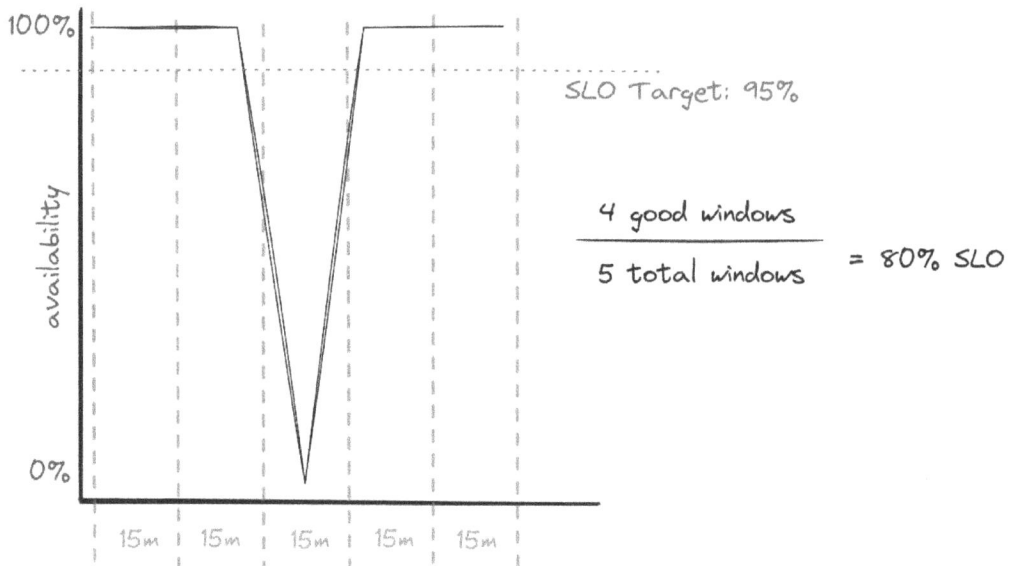

Figure 13.3 – Window-based SLO

This may look undesirable, but consider the following situation. If your service received around 20 requests per minute during all of the good windows, but it spiked up to 500 requests per second from some anomalous traffic during the single bad window, it would take almost 16 days of normal traffic succeeding to offset those bad requests and get back to just 50% availability in a request-based SLO. In contrast, a window-based SLO weighs all time windows equally, so it would only take 15 more good time windows (3.75 hours) to get back to the desired 95% availability.

It may seem like window-based SLOs are clearly the better choice since they are more resilient to outliers and keep your SLO value higher. However, think carefully about which type is the best fit for your services. You will likely end up with a mixture of both types.

A request-based SLO may be more desirable, especially in cases where failed requests equal lost revenue (e.g., failing requests to an API endpoint to provide payment on an online shopping service). In contrast, consider Prometheus itself; it may make more sense to use a window-based SLO for measuring Prometheus's availability since you'll tend to care more about the length of time Prometheus is down as opposed to how many failed requests there were to Prometheus while it was down.

Now that we have some foundational knowledge of the purpose of SLOs and their types, we can put it into practice by seeing how we can leverage Prometheus metrics and SLIs and define SLOs using them.

Defining SLOs with Prometheus data

Prometheus metrics are great for SLIs since they are generally specific enough to pick out individual aspects of a service's performance but still generic enough to be aggregated into an overarching view of how that service is performing. Once you understand the patterns of defining SLOs with Prometheus, the most challenging part is really just identifying which of your hundreds of thousands (or millions) of metrics are best suited to being the SLIs for SLOs.

Once you have your SLI(s) picked out, the SLOs in Prometheus are typically defined in recording rules. Since SLIs will normally need to be aggregated across multiple instances or dimensions, it's simpler to precompute their values instead of running their queries over long time ranges. I also recommend keeping the naming of your SLO recording rules consistent so that it is easy and intuitive to query for SLO metrics across services, for example, using the same resultant series name but with different label values.

Take the following Prometheus recording rule as an example:

```
groups:
  - name: slo-recording-rules-prometheus
    rules:
      - record: slo:sli_success:ratio5m
        expr: |
          clamp_min(
            sum without (pod, instance) (
              increase(prometheus_http_request_duration_seconds_
bucket{le="0.1"}[5m])
            ),
            1
          )
        / ignoring (le)
          clamp_min(
            sum without (pod, instance) (
              increase(prometheus_http_request_duration_seconds_
bucket{le="+Inf"}[5m])
            ),
            1
          )
        labels:
          service: prometheus
          slo: prometheus-request-latency
          window: 5m
```

This recording rule creates a new time series, named slo:sli_success:ratio5m, with labels that help identify the time series that the SLO is related to. Then, for other SLOs, we would create recording rules that use the same time series name but different labels.

Measuring an SLO only over a 5-minute window isn't particularly helpful, though, so we could add another recording rule to measure the value of our existing recording rule over 30 days:

```
- record: slo:sli_success:ratio30d
  expr: |
      sum_over_time(slo:sli_success:ratio5m{service="prometheus",
slo="prometheus-request-latency"}[30d])
      / ignoring (window)
      count_over_time(slo:sli_
success:ratio5m{service="prometheus", slo="prometheus-request-
latency"}[30d])
    labels:
      service: prometheus
      slo: prometheus-request-latency
      window: 30d
```

This new recording rule takes the average of our previous rule over a 30-day time period. Notably, it does not use the `avg_over_time` function. This is to be consistent with popular Prometheus SLO tools, such as Sloth and Pyrra, which we'll look at later.

With that recording rule in place, we could query `slo:sli_success:ratio30d` to see what our SLO measurement is for the last 30 days of data. However, what if we wanted this to be a window-based SLO? Let's see how the same query can be slightly modified to transform this request-based SLO to a window-based version.

Window-based SLOs

Window-based SLOs can be created in Prometheus by leveraging **subqueries** (see *Chapter 3*) and the PromQL `bool` keyword.

The `bool` keyword works by returning a `0` or a `1` depending on whether or not a threshold is met using a comparison operator such as `>=` and `<`. In this case, the threshold is the threshold that determines whether or not a window should be considered to be good.

The time range for the subquery is in the format `[SLO_TimeFrame:WindowSize]`, so to create our SLI with a window of 5 minutes and measure it over 30 days for our SLO, we would use a range selector of `[30d:5m]`. This time range is used to wrap the whole query, including the `bool` part, and then an average of the results over time is taken.

Continuing with our same example but transforming it to be a window-based SLO, the recording rule looks like the following:

```
- record: slo:sli_success:ratio30d
  expr: |
    avg_over_time(
      (
```

```
                    (
                        clamp_min(
                            sum without (pod, instance) (
                                increase(prometheus_http_request_duration_
  seconds_bucket{le="0.1"}[5m])
                            ),
                            1
                        )
                    / ignoring (le)
                        clamp_min(
                            sum without (pod, instance) (
                                increase(prometheus_http_request_duration_
  seconds_bucket{le="+Inf"}[5m])
                            ),
                            1
                        )
                    )
                >= bool 0.999
            ) [30d:5m]
        )
    labels:
      service: prometheus
      slo: prometheus-request-latency
      window: 30d
      sli_window: 5m
```

In this example, we set a success threshold of 99.9% for determining whether or not a window is a good window. Since the `bool` keyword ensures the values of our SLI query will always be a `1` or a `0`, we can average those values to get the percentage for our SLO.

Using recording rules to measure SLOs over time is foundational to implementing SLOs with Prometheus data, but you probably still want to be alerted if your SLO is in danger of not being met, right? Let's talk about how that is done.

Alerting on SLOs

Alerting on SLOs has less to do with alerting on the SLO being below its target and more to do with it **trending** towards breaching its target. After all, if you only alert on your SLO being below its target, it'll always be too late to do anything about it. Instead, alerting focuses on how quickly you are consuming your **error budget**.

Your error budget is the difference between your SLO target and 100%. An SLO target of 99% has an error budget of 1%, an SLO target of 99.9% has an error budget of 0.1%, and so on. The most effective way of alerting on burning through your error budget is through the usage of **multi-window, multi-burn rate** alerts.

Multi-window, multi-burn rate alerts entail specifying both a long window and a short window in an alert and only alerting if both are exceeding their burn-rate thresholds. By having a short window in addition to a long one, we can ensure our alerts resolve more quickly once we stop burning through our error budget.

Burn rates, in this case, are how quickly an SLO's error budget is being consumed. The commonly used thresholds are based on this table, the values of which come from *Chapter 5* in the Google SRE Workbook:

Short Window	Long Window	Burn Rate	Error Budget Consumption
5m	1h	14.4	2%
30m	6h	6	5%
2h	24h	3	10%
6h	3d	1	10%

Table 13.1 – Multi-window, multi-burn rate values

In a Prometheus alert, the rapid consumption of an error budget (the top two rows in the table) may look like the following:

```
- alert: SLOBurnRateHigh
  expr: |
    (
        slo:sli_success:ratio5m{slo="prometheus-request-
latency"} > (14.4 * 0.001)
        and
        slo:sli_success:ratio1h{slo="prometheus-request-
latency"} > (14.4 * 0.001)
    )
    or
    (
        slo:sli_success:ratio30m{slo="prometheus-request-
latency"} > (6 * 0.001)
        and
        slo:sli_success:ratio6h{slo="prometheus-request-
latency"} > (6 * 0.001)
    )
  labels:
    service: prometheus
    slo: prometheus-request-latency
```

These sorts of alerts can quickly get tedious to write by hand, especially with all the recording rules and various windows involved in making it happen. Surely there's a better way to do this, right? As it so happens, there are multiple better ways. So, let's talk about two: Sloth and Pyrra.

Using Sloth and Pyrra for SLOs

Sloth (`https://github.com/slok/sloth`) and **Pyrra** (`https://github.com/pyrra-dev/pyrra`) are two open source projects that are both designed to assist in the definition of SLOs using Prometheus data and the generation of the requisite recording and alerting rules for them. Both projects work by abstracting away the direct definition of recording and alerting rules and instead using their own YAML formats to help define SLOs.

Both projects have been around since 2021, so neither has a clear edge in terms of maturity. Pyrra is still under more active development than Sloth, as Sloth's maintainer has stated that they consider Sloth to be more-or-less in its final form and don't plan on adding new features. Take that as you will, but either project is a good and valid choice on your Prometheus SLO journey. Consequently, we'll look at the basics of both.

Sloth

Sloth works by generating the appropriate recording rules and alerting rules for your SLOs and writing them out to a file for you to deploy with the rest of your Prometheus rules. An added convenience feature of Sloth is that it also creates recording rules for your SLO target, which provides you with a metric you can query to see what your goal is without needing to consult the actual Prometheus rules files.

Sloth's default mode for defining SLOs is focused on request-based SLOs, so we'll build one of those first. Consequently, we need to identify two SLIs to put into our configuration: the SLI representing the errors for our SLO and the SLI representing the total requests.

By continuing our example using the `prometheus_http_request_duration_seconds_bucket` metric, we can build out our request-based SLO using Sloth. The YAML file we'll pass to Sloth looks like the following:

```yaml
version: "prometheus/v1"
service: "prometheus"
slos:
  - name: "prometheus-request-latency"
    objective: 99.9
    description: "SLO based on % of HTTP request responded to in
<=100ms."
    sli:
      events:
        error_query: |
            sum without (pod, instance, le) (
              increase(prometheus_http_request_duration_seconds_
bucket{le="+Inf"}[{{.window}}])
            ) - ignoring (le)
            sum without (pod, instance, le) (
```

```
                    increase(prometheus_http_request_duration_seconds_
   bucket{le="0.1"}[{{.window}}])
                )
          total_query: |
             clamp_min(
                sum without (pod, instance, le) (
                    increase(prometheus_http_request_duration_seconds_
   bucket{le="+Inf"}[{{.window}}])
                ),
                1
             )
       alerting:
         name: PrometheusHighRequestLatency
         labels:
           category: "latency"
         annotations:
           summary: "Prometheus is not serving >=99.9% of requests in
   <100ms."
         page_alert:
           labels:
             severity: critical
         ticket_alert:
           labels:
             severity: warning
```

This accomplishes the same thing we did manually earlier but with a few tweaks. First, you may notice that we don't specify a time window directly. Instead, we use {{.window}} as the range selector value. This is how Sloth uses the Go template language syntax to automatically replace that placeholder value as it templates out its numerous recording rules.

Second, you may notice that we had to change the numerator in our ratio to be representative of the number of requests that didn't meet our threshold (errors instead of successes). We can accomplish that by subtracting the number of requests that did meet the threshold from the total number of requests.

Finally, there is the alerting section. It may not be intuitive to understand the distinction between a page alert and a ticket alert since the names are a bit arbitrary. Essentially, a page alert is a critical alert for when your SLO's error budget is being burned through fast and needs immediate attention. On the other hand, a ticket alert is more of a warning that your error budget is on track to eventually be depleted before the end of the SLO's measurement window, but you have plenty of time to fix it. Either type of alert can be disabled by setting disable: true inside of its YAML block.

Window-based SLOs in Sloth

If you want to use a window-based SLO instead of a request-based SLO in Sloth, you need to provide the raw SLI instead of the error_query and total_query fields. This can still leverage the {{.window}} placeholder, but your query will go under a field called raw (instead of events) and use the error_ratio_query key under raw. The SLI query would essentially be the same as what we did for our window-based SLO earlier but with the same modifications to produce an error ratio instead of a success ratio.

With the configuration explained, let's see what we actually get out of this. We can run Sloth's generate subcommand to emit Prometheus rules. It looks like the following:

```
sloth generate -i mastering-prometheus/ch13/sloth.yaml \
    -o mastering-prometheus/ch13/sloth-generated.yaml
```

The output file then looks like the following:

```
groups:
- name: sloth-slo-sli-recordings-prometheus-prometheus-request-latency
  rules:
  - record: slo:sli_error:ratio_rate5m
    expr: |
      (sum without (pod, instance, le) (
         increase(prometheus_http_request_duration_seconds_
bucket{le="+Inf"}[5m])
      ) - ignoring (le)
      sum without (pod, instance, le) (
         increase(prometheus_http_request_duration_seconds_
bucket{le="0.1"}[5m])
      )
      )
      /
      (clamp_min(
        sum without (pod, instance, le) (
           increase(prometheus_http_request_duration_seconds_
bucket{le="+Inf"}[5m])
        ),
        1
      )
      )
    labels:
      sloth_id: prometheus-prometheus-request-latency
      sloth_service: prometheus
      sloth_slo: prometheus-request-latency
      sloth_window: 5m
```

The above looks very similar to what we already wrote by hand, but Sloth takes the guesswork out of it and templates out all of the recording rules and alerting rules we need across all of our windows. Since Sloth outputs a whopping 17 different Prometheus rules across three different rule groups, the full generated output is omitted from this book. However, you can consult the folder for this chapter in this book's GitHub repository to see the fully generated output or try running Sloth for yourself!

This is just a basic overview of how Sloth works, so I encourage you to explore the project's documentation for a more in-depth look at all it can do. With our understanding of how Sloth works, let's see how Pyrra works, too.

Pyrra

When you first look at the GitHub project page for Pyrra, you'll notice that it has one killer feature that Sloth lacks: a shiny web UI. The web UI is solely for visualizing your SLOs, though. They are still defined in YAML.

Another difference is the expected deployment mode of Pyrra. Whereas Sloth expects you to run it each time you want to generate updated Prometheus rules for your SLOs, Pyrra is designed to be run as a long-living process that is co-located with your Prometheus instance. If you run multiple Prometheus instances, the authors recommend running one Pyrra instance for each Prometheus instance you're running.

Nevertheless, for this chapter, we're mainly interested in how Pyrra works and not deploying it to our Kubernetes cluster. Thankfully, Pyrra added support for a similar `generate` command regarding its CLI tool, and we can use it to see what its output looks like. Pyrra's YAML format is somewhat close to Sloth's and should be easy to pick up if you're familiar with Sloth already. One slight difference is that since Pyrra was originally designed to be run in Kubernetes, its configuration file format is in the style of a Kubernetes **custom resource definition** (**CRD**).

The Pyrra version of the same SLO we defined with Sloth looks like the following:

```
apiVersion: pyrra.dev/v1alpha1
kind: ServiceLevelObjective
metadata:
  name: prometheus-request-latency
  namespace: prometheus
spec:
  target: "99.9"
  window: 30d
  description: Percentage of requests Prometheus is fulfilling in
<100ms.
  indicator:
    latency:
      success:
```

```
      metric: prometheus_http_request_duration_seconds_
bucket{le="0.1"}
    total:
      metric: prometheus_http_request_duration_seconds_count
    grouping:
      - handler
  alerting:
    name: PrometheusHighRequestLatency
```

When generating that with `pyrra generate`, we end up with an output that is roughly similar to that of Sloth. Similar to Sloth, it generates over a dozen rules, so the full output isn't included in this book but is available in the folder for this chapter in the book's GitHub repository. However, as an example, one of the rules that was output looks like the following:

```
  - expr: (sum by (handler) (rate(prometheus_http_request_duration_
seconds_count[5m]))
      - sum by (handler) (rate(prometheus_http_request_duration_
seconds_bucket{le="0.1"}[5m])))
      / sum by (handler) (rate(prometheus_http_request_duration_
seconds_count[5m]))
    labels:
      slo: prometheus-request-latency
    record: prometheus_http_request_duration_seconds:burnrate5m
```

Unfortunately, we had to give up our `clamp_min` safeguards because Pyrra only works with vector selectors, and you cannot use PromQL functions in the `metric` fields. Additionally, for the same reason, we lost out on some labels that we had previously preserved using `without` in our query. Instead, Pyrra has a `grouping` field that is used to specify labels to be included in a `by` aggregation.

If you review the full recording rules that are output, you'll also see that the short and long windows don't quite align nicely. Whereas Sloth uses windows of 5m, 30m, 1h, 2h, 6h, 1d, 3d, and 30d, Pyrra generated windows of 5m, 32m, 1h4m, 2h9m, 6h26m, 1d1h43m, and 4d6h51m. This is because – at the time of writing – Pyrra hardcodes burn rate window calculations assuming you're using a 28-day sliding window for your SLOs, as opposed to 30 days, like with Sloth. This isn't a big deal, but it certainly will raise questions if you try to get other teams in your organization to adopt Pyrra for their SLOs without adjusting windows.

On the bright side, Pyrra does have native support for our latency SLO type, so we didn't have to modify any queries to derive the error rate – it did that for us.

Overall, both Pyrra and Sloth accomplish the same tasks with slightly different details involved. I'd recommend trying them both out to see which is the most intuitive to you and go with that. After all, a tool is just a tool. The SLOs you get at the end are what really matter, and both of these tools help you get there.

Summary

In this chapter, we went over the basics of SLOs so that we could see how Prometheus can help enable the definition and adoption of SLOs in your organization. We also explored and experimented with two popular open-source tools – Sloth and Pyrra – that help make SLOs easy when using Prometheus by implementing best practices with minimal code involved. For example, we wrote >80% fewer lines of code using Pyrra than we would have if we wrote all the Prometheus rules by hand.

In our next (and penultimate) chapter, we'll look at one of the most exciting projects in the observability space today: OpenTelemetry. We'll see how OpenTelemetry seeks to unify observability signals and how Prometheus is able to be a part of that unification through both official support in Prometheus itself and extensions in the OpenTelemetry project's code bases.

Further reading

To learn more about the topics that were covered in this chapter, take a look at the following resources:

- *SLOs*:

  - *"Implementing SLOs," Chapter 2 from The Site Reliability Workbook*: `https://sre.google/workbook/implementing-slos/`

  - *"Alerting on SLOs," Chapter 5 from The Site Reliability Workbook* `https://sre.google/workbook/alerting-on-slos/`

  - *Touching Grass with SLOs*: `https://www.honeycomb.io/blog/honeycomb-internal-slo`

  - *Alerting on SLOs like Pros*: `https://developers.soundcloud.com/blog/alerting-on-slos`

  - *A practical guide for implementing SLOs*: `https://last9.io/blog/a-practical-guide-to-implementing-slos/`

- *Sloth*:

  - `https://sloth.dev/`

  - *Considered to be in its final form*: `https://github.com/slok/sloth/issues/521`

- *Pyrra*: `https://github.com/pyrra-dev/pyrra`

14

Integrating Prometheus with OpenTelemetry

In my career, I have never encountered an observability project on the level of OpenTelemetry. Not since the introduction of SNMP has there been such a widely adopted effort to standardize around a data format and ecosystem for application and system telemetry.

To many a cynical modern mind, the concepts of vendor agnosticism and unifying standards may seem like a shining ideal whose light dwindled sometime in the late 20th century after common Internet standards were established around things such as HTTP and DNS. Sure, there are updates and iterations on those things, but what entirely new technology standard really sees widespread adoption? So much modern innovation is proprietary and vendor-specific, perpetuating cycles of vendor lock-in. Yet, some of the worst perpetrators of vendor lock-in – cloud observability and **application performance monitoring** (**APM**) vendors – have forsaken their proprietary protocols and data formats to embrace OpenTelemetry instead.

Maybe you have heard of OpenTelemetry already and have a solid understanding of it. Great! This chapter is for you. Or perhaps you've never heard of OpenTelemetry until two paragraphs ago or have heard of it but don't really know much about it besides the name. Great! This chapter is for you, too.

We're going to go over the basics of OpenTelemetry to understand what it is, the problems it solves, and the ecosystem around it. Then, we'll see how it ties into Prometheus and can be a part of your wider Prometheus environment. If you already feel like you have a solid grasp of OpenTelemetry fundamentals, feel free to skip ahead to the Prometheus-specific sections. Still, with that in mind, we'll be covering these main topics:

- Introducing OpenTelemetry
- Collecting Prometheus metrics with the OpenTelemetry Collector
- Sending metrics to Prometheus with the OpenTelemetry Collector

Let's get started!

Technical requirements

For this chapter, we'll be leveraging our Prometheus environment built in *Chapter 2*. In addition to that, you'll need these tools installed to follow along:

- **OpenTelemetry Collector** *(otelcol; not otelcol-contrib)*:`https://github.com/open-telemetry/opentelemetry-collector-releases/releases/tag/v0.92.0`

The code used in this chapter is available at `https://github.com/PacktPublishing/Mastering-Prometheus`.

Introducing OpenTelemetry

At its core, **OpenTelemetry** (commonly abbreviated as **OTel**) is not a tangible technology such as Prometheus, Thanos, or Kubernetes. It's not something that you "run," per se. Instead, OpenTelemetry is a *specification*.

The OpenTelemetry project itself was born out of a rare event: the consolidation of two competing open source projects. Before OpenTelemetry became a project, two competing open source standards sought to address the problem of vendor lock-in in the observability space: **OpenCensus** and **OpenTracing**.

OpenCensus, which began in 2017, was the open source solution sponsored by Google and covered both tracing and metrics. OpenTracing, which began in 2016, was a CNCF project (like Prometheus and Kubernetes) and was – as its name implies – primarily focused on tracing. However, as one of my favorite XKCD comics explains, multiple competing standards are less than ideal.

Figure 14.1 – XKCD 927, "Standards"(Source: https://xkcd.com/927)

In a rare exception to this phenomenon of competing standards, the teams from both OpenCensus and OpenTracing worked together to form the OpenTelemetry project under the CNCF umbrella in 2019. Since then, hundreds of contributors have worked together to develop OpenTelemetry into a project that meets and exceeds the prior functionality of both projects. The primary way that this was accomplished was through the development of a formal specification that guides and governs the usage and implementation of OpenTelemetry.

OTel specification

The specification for OpenTelemetry is at version 1.29.0 at the time of writing and is accessible online at `https://opentelemetry.io/docs/specs/otel/`. It consists of multiple subspecifications for the OpenTelemetry API, associated development SDKs, and semantic conventions for standardizing things such as the naming of metadata (also referred to as `context`) fields.

OpenTelemetry's API is organized around the three primary observability signals we discussed way back in *Chapter 1*: metrics, logs, and traces. It defines how each of those signals should be used and how to interface with them. For example, how to instantiate and update telemetry data within code bases. In turn, the various programming language-specific SDKs each implement the APIs in consistent ways to ensure the portability of the concepts and expected behavior across languages.

The API and SDKs are really more concerned with how you instrument your code to emit data, though. It's somewhat akin to a Prometheus client library, such as `client_golang`, which we used in *Chapter 8* to expose metrics in a simple exporter:

OTel specification	Target audience
API	OpenTelemetry developers creating and maintaining SDKs.
SDKs	Application developers instrumenting their code to emit telemetry data in a consistent way that conforms with OpenTelemetry expectations.
Semantic conventions	Both producers and consumers of telemetry data conforming to OpenTelemetry's specifications. Guides producers in ensuring consistent naming and values for telemetry and its attached metadata and ensures the predictable querying and consumption of that data by consumers.

Table 14.1 – Target audiences for parts of the OTel specification

As an SRE, the target audience for the APIs and SDKs isn't really me, and it most likely isn't you, either. Instead, my favorite part of the OpenTelemetry specification is its semantic conventions.

> **But what about the SDKs?**
>
> The OpenTelemetry SDKs provide the ability to emit traces, metrics, and even logs. While you can certainly use OTel libraries for your metrics, it's still common to use Prometheus's client libraries for your Prometheus metrics. This is due to some differences in how OTel defines metrics vs. Prometheus. For example, in OTel, a counter metric does not have to be monotonous.

OpenTelemetry's semantic conventions lay out all sorts of conventions surrounding the way that metrics, logs, and traces should be formatted and named, what metadata should be attached to them, what the metadata keys should be named, and more in extensive detail. It solves one of the most frustrating things about using Prometheus across many teams and services: inconsistency in things such as metric and label naming. Wouldn't it be grand if teams all exposed their service's HTTP metrics using the same metric name and properly used labels to distinguish them from other services instead of using the metric name to do so? One can dream.

> **Semantic Conventions**
>
> The full collection of OpenTelemetry's semantic conventions grew so large that it now has its own GitHub repository. To explore the various defined conventions, check out their latest versions at `https://github.com/open-telemetry/semantic-conventions/`.

That covers the idea of how OpenTelemetry standardizes how telemetry data is defined, formatted, and exposed. However, there's still more to OpenTelemetry in how it standardizes the **transmission** of this telemetry. For that, there is a separate specification for the **OpenTelemetry line protocol** (**OTLP**).

OpenTelemetry line protocol

The OTLP is what allows the interoperability of OTel to shine. Providing a standardized way for OpenTelemetry data to be passed around between telemetry producers and their associated backends is at the heart of enabling OTel's goal of providing vendor neutrality.

Many observability SaaS providers, such as New Relic, have already transitioned to having OTLP be the preferred way that telemetry data is sent to them. Since the format of the data is consistent, you – as a user – can simply point your data being sent over OTLP to a different backend destination, most likely without having to make any code changes.

Additionally, OTLP supports all three primary observability signals, so you don't need to be concerned with whether you are sending logs, metrics, or traces to a backend; this can be left up to the server receiving OTLP data to sort out during the parsing of the received data based on the OTLP specification.

OTLP specification

The OTLP Specification is a distinct, separate specification from the primary OpenTelemetry specification. At the time of writing, the OTLP specification is at version 1.1.0. Its purpose is to describe the way that telemetry data should be encoded, sent, and received between telemetry sources and their destinations.

As users – rather than implementers – of OTLP, we only need to be concerned with the two ways that OTLP is implemented. The OTLP spec provides two ways that data can be sent over OTLP: via gRPC and via HTTP. If you're in doubt about which one to choose, I recommend gRPC. It was the original implementation of OTLP (support for the HTTP transport came later) and, therefore, tends to be more common and preferred.

> **It's HTTP all the way down**
>
> If you're familiar with gRPC, you may know that gRPC already technically operates over HTTP but only on servers that support at least HTTP/2. Consequently, when you see or hear "OTLP over HTTP," know that it generally refers to sending HTTP data over HTTP/1.1 (usually for compatibility reasons) instead of using gRPC.

The usage of OTLP will come into play later when we send data directly to Prometheus using OTLP. But how do we send that data if we know that Prometheus is based on a "pull" model? Something else has to pull that data to send it to Prometheus, and that "something" we'll use is the OpenTelemetry collector.

OpenTelemetry collector

The **OpenTelemetry collector** (often abbreviated as **otelcol**) is the one piece of the OpenTelemetry project that you can actually run. Its code base and published releases are located at `https://github.com/open-telemetry/opentelemetry-collector`. It is an extensible agent meant to act as an intermediary between telemetry producers and storage backends for various types of telemetry. It provides a way to receive telemetry data from one or more services, make modifications to it, and send it off to somewhere else that receives the telemetry data.

Each different telemetry type gets its own **pipeline**, which is comprised of three primary component types in the collector's configuration: receivers, processors, and exporters.

Receivers

Receivers are the input of the OpenTelemetry collector. They configure how data is received when it first enters the collector. These receivers can either be active (pull-based) or passive (push-based). For example, an application could push data to an OTLP receiver configured on the collector, or the collector itself could actively scrape metric data from an application in the same way that Prometheus does.

Processors

Processors are the middleware of the collector and perform the filtering and transformation of received data. Processors are optional, so you do not need to configure any between a receiver and an exporter. However, they are a helpful tool for doing things such as adding common metadata to all received telemetry or filtering and combining telemetry. They tend to serve similar purposes to what Prometheus's `metric_relabel_configs` accomplish – transforming and filtering data that were already received prior to it finally being stored or sent elsewhere.

Exporters

Exporters are the final stage in the pipeline for telemetry entering the OpenTelemetry collector. They configure what data is sent to where after it has been received and processed, for example, sending logs to your log storage backend or even pushing metrics to a Prometheus instance.

If you want to send or receive data from Prometheus, though, you'll notice that the main OpenTelemetry Collector repository linked previously does not seem to have any way to do that. What gives? Well, in order to have access to all of the available integrations with the OpenTelemetry collector, we need to instead look at the OpenTelemetry collector contrib repository.

OpenTelemetry collector contrib

Whereas the main OpenTelemetry collector repository is extremely barebones in the functionality it provides, the OpenTelemetry collector contrib repository (`https://github.com/open-telemetry/opentelemetry-collector-contrib/`) has dozens and dozens of available plugins for receivers, processors, and exporters.

The reasoning behind the separate repositories is to reduce bloat in the binaries produced by the collector. Since there are so many plugins available for use in the contrib repository, its releases tend to be much larger. A minimal subset of components from the contrib repository, including the Prometheus receiver and Prometheus exporter, are also included in the "core" releases of the OpenTelemetry collector. Consequently, using the standard `otelcol` release (i.e., the one without a `-contrib` suffix) is acceptable for the purposes of this chapter.

> **Isn't there a middle ground?**
>
> Rather than having to choose between an "all or nothing" approach when deciding whether you should run `otelcol` or `otelcol-contrib`, OpenTelemetry provides a method to build your own version of the collector that includes only the components you want. The OpenTelemetry collector builder is located at `https://github.com/open-telemetry/opentelemetry-collector/tree/main/cmd/builder`.
>
> The usage of the builder is more targeted at advanced, mature deployments of OpenTelemetry. When starting out with the OpenTelemetry collector, using the `otelcol-contrib` releases is recommended to ensure you have access to all components while experimenting and building out your telemetry pipelines.

Now that we have a basic understanding of how the OpenTelemetry collector works and is structured, let's see how to start integrating it with Prometheus.

Collecting Prometheus metrics with the OpenTelemetry collector

The first step in integrating the OpenTelemetry collector with Prometheus is to get metric data. There are myriad ways to get metric data into the OpenTelemetry collector, but we'll operate based on the assumption that we're collecting data from things that already expose Prometheus metrics. In order to collect data from those applications, we can use the `prometheus` receiver from the OpenTelemetry collector contrib repository.

The `prometheus` receiver supports all of the same configuration settings that Prometheus itself supports within its `scrape_configs` section (including service discovery!). Consequently, if you've made it this far, it should be easy to configure.

To keep things simple, we'll just configure the OpenTelemetry collector to scrape itself. Rather than sending data to Prometheus just yet, we'll use the `debug` exporter to output data directly to our terminal instead.

The configuration to accomplish that is stored in this chapter's folder in this book's GitHub repo. It is named `otelcol-config-debug.yaml` and looks like the following:

```yaml
receivers:
  prometheus:
    config:
      scrape_configs:
        - job_name: "otel-collector"
          scrape_interval: 15s
          static_configs:
            - targets: ["127.0.0.1:8888"]

exporters:
  debug:
    verbosity: detailed

service:
  pipelines:
    metrics:
      receivers: [prometheus]
      exporters: [debug]
```

Since the OpenTelemetry collector automatically exposes Prometheus metrics about itself on port 8888 using the standard Prometheus URL path of /metrics, we don't have to do any additional configuration of the collector itself. To see this in action, we can run otelcol using this config file:

```
$ otelcol --config=mastering-prometheus/ch14/otelcol-config-debug.yaml
```

After a few seconds, the first scrape cycle will run, and you'll see a dump of all of the metrics that looks like the following:

```
Metric #9
Descriptor:
     -> Name: otelcol_process_runtime_heap_alloc_bytes
     -> Description: Bytes of allocated heap objects (see 'go doc
runtime.MemStats.HeapAlloc')
     -> Unit:
     -> DataType: Gauge
NumberDataPoints #0
Data point attributes:
     -> service_instance_id: Str(faef9568-6c7e-41c4-bfcd-2788234091de)
     -> service_name: Str(otelcol)
     -> service_version: Str(0.92.0)
StartTimestamp: 1970-01-01 00:00:00 +0000 UTC
Timestamp: 2024-01-20 19:47:18.355 +0000 UTC
Value: 9899336.000000
```

This gives great insight into how Prometheus metrics are represented internally within the OpenTelemetry collector. OpenTelemetry represents some metric data in different ways than Prometheus. There is plenty of overlap – such as how this "Gauge" metric maps to a Prometheus gauge. However, some other metric types may be different. For example, a Prometheus "counter" will be represented as a "Sum" in the OTel data model.

Most of the other fields should be self-explanatory based on your understanding of Prometheus already, but it's worth drawing attention to the Data point attributes field. The values within that field are what map to Prometheus labels. Consequently, each key we see there will be mapped to a label name when we send this data to Prometheus. Speaking of which, let's see how we can do that.

Sending metrics to Prometheus with the OpenTelemetry collector

At the time of writing, sending data to Prometheus over OTLP is still considered an experimental feature. In fact, the functionality came out during the process of writing this book (in Prometheus version 2.47.0), so we'll need to upgrade the version of Prometheus we've been running so far in order to enable support for it.

Configuring Prometheus

To support the OTLP receiver on Prometheus, we need to adjust our Helm chart to install a newer version of Prometheus and add a `--enable-feature=otlp-write-receiver` flag to the Prometheus process. Additionally, to avoid any complications inherent in pushing data to multiple Prometheus replicas, we'll scale back down to a single replica.

The Helm values file we'll use looks like the following:

```
grafana:
  enabled: true
  defaultDashboardsTimezone: browser
  adminUser: root
  adminPassword: m@ster1ngPr0m3th3us
prometheus:
  prometheusSpec:
    serviceMonitorSelectorNilUsesHelmValues: false
    enableFeatures:
      - otlp-write-receiver
    image:
      tag: v2.49.1
  cleanPrometheusOperatorObjectNames: true
```

We can deploy that by running this Helm command:

```
$ helm upgrade --namespace prometheus \
    --version 47.0.0 \
    --values mastering-prometheus/ch14/prom-values.yaml \
    mastering-prometheus \
    prometheus-community/kube-prometheus-stack
```

Once that is finished deploying, we can test sending data from the OpenTelemetry collector to it.

In order to forward the data, we'll need to make sure that Prometheus is accessible from our computer where OpenTelemetry collector is running. So, go ahead and port-forward to your Prometheus instance locally:

```
$ kubectl port-forward svc/mastering-prometheus-kube-prometheus
9090:9090
```

While that's running in the background, let's look at what changes we need to make to our OpenTelemetry collector configuration to start sending to Prometheus.

Configuring OpenTelemetry collector

To send our metric data to Prometheus from the collector, we only need to change the exporter we are using in the collector configuration. Since we're sending data over OTLP, and since Prometheus's API is HTTP-based, we can use the `otlphttp` exporter. The receiver section of our configuration stays the same, but now our exporter and pipeline sections look like the following:

```
exporters:
  otlphttp:
    endpoint: http://localhost:9090/api/v1/otlp

service:
  pipelines:
    metrics:
      receivers: [prometheus]
      exporters: [otlphttp]
```

The `otlphttp` exporter automatically adds relevant suffixes to URLs (for example, `/v1/metrics`, `/v1/logs`, etc.), so we only need to specify `/api/v1/otlp` even though the complete Prometheus API endpoint is `/api/v1/otlp/v1/metrics`.

That updated configuration is in a file in our GitHub repository and is named `otelcol-config-prom.yaml`. When running it, we should see the metrics about the OpenTelemetry collector begin showing up in our Prometheus instance:

```
$ otelcol --config=mastering-prometheus/ch14/otelcol-config-prom.yaml
```

Since we're already port-forwarding to our Prometheus instance, we can go to `http://localhost:9090` in our web browser and query for metrics from our collector to confirm:

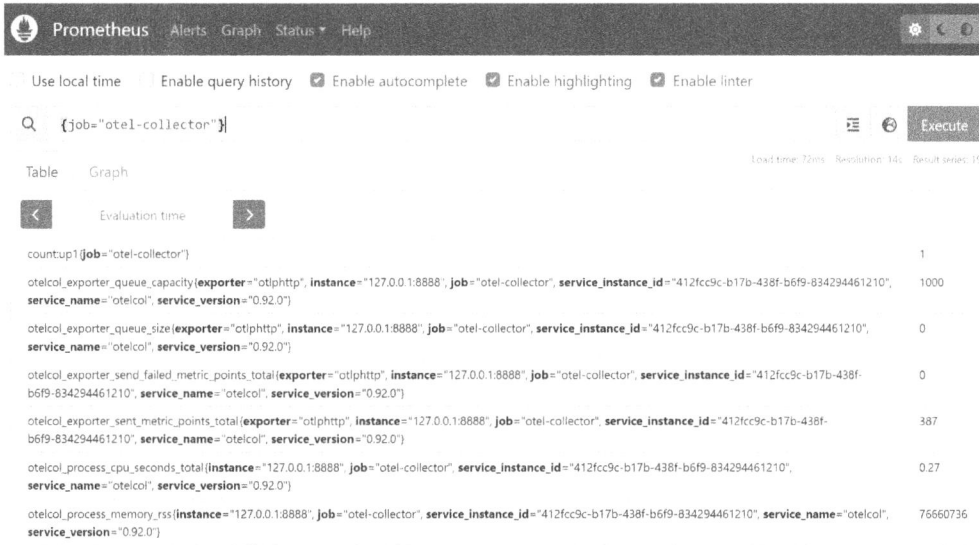

Figure 14.2 – OpenTelemetry collector metrics pushed via OTLP

While this is undoubtedly very cool and exciting, I do have some words of caution. Don't be fooled by the presence of an up metric. That metric is coming from the prometheus receiver on the collector when it scrapes the collector's metrics. Consequently, please don't rely on it for alerting to ensure that the process of sending data is actually succeeding.

In a real-world deployment, even if you configured your OpenTelemetry collector to push data to Prometheus, you would at least want your Prometheus instance to be the one directly scraping metrics from the collector process to reliably know if the collector is up and if it's succeeding in pushing data to your Prometheus (see the otelcol_exporter_send_failed_metric_points_total metric). Of course, at that point of mixing both pull-based and push-based models, you really have to consider if it's not worth using the pull-based model for everything. Even with its new OTLP receiver functionality, Prometheus is still designed to be used as a pull-based system, and any deviation from that should be for a good reason.

Summary

In this chapter, we learned about the OpenTelemetry project and the exciting prospects it poses for the future of observability. Rather than superseding Prometheus or rendering it obsolete, we saw how the two are able to natively integrate and provide added functionality.

Importantly, we saw that OpenTelemetry's integration with Prometheus should be used with intentionality and not necessarily as a wholesale replacement of Prometheus's default, direct pull-based model. Nevertheless, OpenTelemetry is an exciting, rapidly developing project that you should keep your eye on for years to come as its integrations develop and improve.

In our next and final chapter, we'll look at what you can do right now to make your observability environment as robust as possible. Remember, way back in *Chapter 1*, when we talked about how Prometheus is just one piece in a larger picture of making a system observable? In our next chapter, we'll look at what the other pieces of that picture look like so that we can go beyond Prometheus to achieve more observable systems.

Further reading

To learn more about the topics that were covered in this chapter, take a look at the following resources:

- *OpenTelemetry specification*: `https://opentelemetry.io/docs/specs/otel/`
- *OpenTelemetry line protocol specification*: `https://opentelemetry.io/docs/specs/otlp/`
- *OpenTelemetry semantic conventions*: `https://opentelemetry.io/docs/specs/semconv/`
- *Recommended books on OpenTelemetry*:
  - Boten, Alex. *Cloud-Native Observability With OpenTelemetry: Learn to Gain Visibility Into Systems by Combining Tracing, Metrics, and Logging With OpenTelemetry*. Packt Publishing, 2022.
  - Parker, Austin, and Ted Young. *Learning OpenTelemetry: Setting up and Operating a Modern Observability System*. O'Reilly Media, 2024.
  - Majors, Charity, et al. *Observability Engineering*. O'Reilly Media, 2022.

## 15

Beyond Prometheus

If you've made it this far, congratulations! By the power vested in me, I hereby declare you a master of Prometheus. Go forth into the world and observe all the things.

Oh. Are you still here? Did you come back when you remembered that Prometheus by itself doesn't make a system fully observable? Either way, I want to make the most of the rest of our time together to discuss the question, "What next?"

Hopefully, by now, you've come to see how Prometheus (and all metrics systems, by extension) only account for a subset of observability. If you accept the idea that the three core observability signals are metrics, logs, and traces, then solely using Prometheus only gets you one-third of the way toward a fully observable system.

By using Prometheus as our starting point, we need to see what other pieces we can add to make our systems more observable. This chapter will take a fresh look at the observability concepts we discussed in *Chapter 1*, focusing on how logs, traces, and more can all (together with the metrics system we've already built with Prometheus) contribute to a more observable system. Additionally, we'll cover a few open source projects that integrate well with Prometheus to improve observability in a cohesive way.

Consequently, in this chapter, we'll be covering these main topics:

- Extending observability past Prometheus
- Connecting the dots across observability systems

Technical requirements

The code used in this chapter is available at `https://github.com/PacktPublishing/Mastering-Prometheus`.

Extending observability past Prometheus

Setting aside the obvious nuance that "no system is truly and fully observable," let's focus on how we can get as close as possible to a fully observable system. Observability is all about the concept of being able to account for "unknown unknowns." In other words, you shouldn't need to know in advance the various ways that a system can break in order to monitor and observe it effectively.

> **Knowing what you don't know**
>
> The concept of "unknown unknowns" was popularized by former United States Secretary of Defense, *Donald Rumsfeld*. It has to do with the idea of not knowing what you don't know or your ignorance of the extent of your ignorance. In contrast, a "known unknown" would be something that you are aware but do not know. When relating this concept to Prometheus, an "unknown unknown" would be a metric that doesn't exist, but a known unknown would be a metric that exists where its value, understandably, isn't known until query time.

Prometheus gets us part of the way towards making our unknown unknowns become known due to its focus on being able to collect millions of individual time series. However, by its very nature, the metrics need to be defined ahead of time. This means that even though we can add hundreds and thousands of metrics to try to make sure we have as many data points as possible, there will inevitably be things that are not accounted for.

In order to allow for more dynamic telemetry, we need to look beyond Prometheus to other telemetry signals that are better suited to that purpose, namely, logs and traces.

Logs

Logs and I have a complicated relationship. On one hand, logs are by far the most accessible telemetry signal to add and use. They predate the existence of modern computer monitors and harken back to the days of computers literally printing output on paper. Consequently, every developer, SRE, project manager, and other IT professional knows what a log is and what purpose it serves.

On the other hand, logs indisputably have the worst signal-to-noise ratio of our three observability signals. I shudder to think how much time I've spent in my career combing through logs, looking for a needle in a haystack of thousands of log lines to track down why something broke. This makes it extremely difficult to extract meaningful value from logs.

The core problem with logs is that the vast majority of them are **unstructured**. This is because logs are generally meant for human consumption. Consequently, for a human, the ease of reading is prioritized over the ease of parsing for a computer. Structured logging is becoming more common, but I don't think that we'll ever quite hit a sweet spot that perfectly balances human readability with machine parsability. It's becoming more common for applications to support outputting logs in a JSON format, but that is not the most readable. In the worst case, a JSON log is output without whitespace that would aid in human readability. In the best case, the whitespace is there, but then a single log "line" can fill up your whole terminal window.

The closest compromise I've seen achieved with logs is through **logfmt**, which is a logging style popularized originally by Heroku but also by various Go projects. For example, it's the format that Prometheus uses for its logs. It looks like the following:

```
ts=2024-01-27T03:00:01.861Z caller=head.go:1287 level=info
component=tsdb msg="WAL checkpoint complete" first=70 last=71
duration=1.18069471s
```

The structure of using whitespace-separated key-value pairs delimited by an equals (=) sign allows for machine parsing while still allowing for relative ease of reading as a human. However, this still loses human readability vs. just printing a timestamp and "WAL checkpoint complete." Consequently, even moving from unstructured logging to `logfmt` is likely to encounter pushback.

Therein lies the problem. Trying to repurpose logs into an observability signal will inevitably result in pain, frustration, and confusion. It's a signal that is full of tradeoffs that mean that you'll never be able to make everyone happy. Nevertheless, logs are still the easiest signal to add in code, so they'll tend to be the "go-to" signal for most developers. Then, they start to get abused. Why add tracing to our application when we can just output multiple log lines in every function call? Why add metrics when we can just count the number of log lines matching a pattern? If you've had to field these types of questions before, I send my consolations, but you are not alone.

Metrics from logs

Inevitably, engineers want to derive metrics from their logs. The thinking goes, "We already have these logs, so do we really need to add metrics, too?" The short answer is yes. Does this tend to feel duplicative if you're incrementing a metric every time you output a log line? Also, yes. But the additional value is well worth the extra few lines of code.

Extrapolating metrics from log lines should **always** be discouraged. By nature, log lines are far more likely to change over time as additional context is added and formatting is altered. Relying on metrics derived from logs ultimately introduces unnecessary risk for gaps in observability due to parsing errors.

Log lines are inherently more dynamic than metrics and are likely to contain more specific information. This makes pattern parsing to perform extrapolation more fragile and subject to failure or otherwise unpredictable/unreliable behavior. This then adds additional overhead in the long term to both the team(s) producing logs and the team(s) operating the solution that is storing/parsing the logs when those teams have to debug and troubleshoot those issues when they inevitably arise.

Instead, directly adding metrics should always be preferred. This eliminates the risk of a fragile translation layer mucking up your observability signals and keeps the two observability signals logically separated. In an ideal world, our various observability signals should be connected but not **dependent** on each other. If we have issues collecting or parsing logs, we shouldn't also lose out on seeing the metrics of our system. However, they should be connected in the sense that it makes it easy to jump from one signal to another to see an application from different angles at the same point in time.

To log or not to log

At this point, maybe I've succeeded in radicalizing you, and you think we should abandon logs entirely as a relic of a bygone age. Is that actually what I believe? Surprisingly, no. Logs still have their place and all of my frustration with them as an observability signal stems from attempts to extend their usage out of that place.

Logs will always be there for you. Even if your traces disappear from your object storage bucket and your Prometheus environment comes crashing down, you'll still have local logs to fall back on. They are undeniably helpful, and all the more so when they are combined with our other telemetry signals, such as Prometheus metrics.

Being able to see a spike in a Prometheus metric, zoom in on that time period, and jump to the log lines for that service during that time frame is a convenience that I don't ever want to live without again. It makes finding that needle in the haystack of logs that much easier.

Traces

While logs are the oldest core observability signal, traces are undoubtedly the newest. Other signals, such as continuous profiling, are still emerging and aren't quite "core" observability signals yet, so tracing still holds that title of being the newcomer (for now). I would not be surprised if you have yet to actually use a tracing system. Admittedly, it's the signal that I am the least experienced with, too. However, when used correctly, traces are life-changing.

Traces have the best signal-to-noise ratio of the observability signals, especially if **sampling** is applied properly. Trace sampling is the idea of only storing a subset of traces that are emitted. Since there is typically one trace for every request or operation that a service performs, this can be a considerable amount of data to deal with. Traces tend to contain many (sometimes measured in dozens!) child spans within them, so it's easy to have more spans than log lines for a request. Predictably, this can be expensive to store and query.

Sampling allows you to store only a subset of requests, and your criteria for which ones to keep can be highly detailed. For example, you could configure sampling to store 1% of all traces by default but 50% of traces where the request took longer than a certain time and 100% of traces where an error was encountered. This helps filter out the noise to ensure that the traces you have provide a high level of value.

Since sampling is effectively a requirement for high-volume services, traces can't serve as a replacement for the high-level metrics view you get with Prometheus. However, they can help to fill in the details that metrics inherently omit.

With that said, for a genuinely effective observability stack, you need to be able to knit together all three primary observability signals into a cohesive view of a service. Therefore, let's discuss how that can be achieved through other observability systems that can complement Prometheus.

Connecting the dots across observability systems

Prometheus is quickly becoming many people's first introduction to the world of observability technologies. Its popularity and seeming ubiquity have consequently led it to influence many other observability projects that seek to replicate its ease of use and simple data model. This has the added benefit of making it easier to jump between observability services without having to change the filters we use to select data.

The most meaningful way to easily connect the dots across our various observability systems is to ensure consistency in the metadata that is attached to the telemetry coming from that system. In the context of Prometheus, that means that our label keys and values for `ServiceX` should be the same in Prometheus as they are in the rest of the systems we use. For example, if your Prometheus metrics for `ServiceX` have a label `app=servicex` on them, your logs for `ServiceX` should **not** have a label of `app=service-x` on them – they should match.

Prometheus's concept of using labels and label matchers is heavily influencing new observability systems. These systems tend to have query languages that resemble PromQL, which makes jumping between the systems much more straightforward. Two up-and-coming solutions that integrate beautifully with Prometheus to provide our missing signals are Loki for logs and Tempo for traces, both from Grafana Labs.

We won't cover the ins and outs of deploying these systems, but let's at least talk briefly about what use they may be able to provide you.

Logging with Loki

Loki does not shy away from its comparison to Prometheus. In fact, its GitHub repository's description (`https://github.com/grafana/loki`) states that it is "Like Prometheus, but for logs."

This comparison stems from a variety of concepts that it borrows from Prometheus, including the usage of similarly formatted labels. Additionally, Loki incorporates other Prometheus designs, such as the usage of a WAL, a PromQL-inspired query language called LogQL, and even a Prometheus-like TSDB index.

Compared to other popular self-hosted logging solutions, such as the **Elastic Stack** (also known as **ELK**), Loki is also significantly simpler to operate. It can run in a single-binary mode where you only need to run one instance of Loki rather than several different servers. Additionally, it uses object storage for storing its data, so you don't need to worry about having massive, several-terabyte disks attached for storing logs.

> **Licensing woes**
>
> It's worth noting that, like Mimir (see *Chapter 9*), Loki is licensed under AGPLv3. The same applies to Tempo, which we'll discuss in the next section.

It integrates natively with Grafana, as would be expected regarding a product from Grafana Labs. This makes it easy to create dashboards that can show your metrics and your logs for a service side-by-side. My team is currently running Loki at a scale of several hundred thousand log lines per second, and I would certainly recommend it over any self-hosted alternative, especially if you value the ability to interconnect your logs and Prometheus metrics.

Tracing with Tempo

Tempo (`https://github.com/grafana/tempo/`) is another recent open source product from Grafana Labs that is also deeply integrated with Prometheus (and Loki, too). It is focused on providing a scalable tracing solution using designs and values similar to those of other Grafana projects, such as Loki and Mimir (see *Chapter 9*).

> **Looks good to me**
>
> The full suite of Grafana products that cover the three core observability signals is commonly referred to as the **LGTM stack** for Loki, Grafana, Tempo, and Mimir. It's a play-on-words (play-on-acronyms?) for the commonly used "LGTM" (looks good to me) comment on pull request reviews.

Similar to Loki, Tempo also implements a PromQL-inspired query language known as TraceQL. This helps to make it easy to jump between your metrics and traces for an application. However, the easiest way to natively jump between metrics and traces actually comes from a feature built into Prometheus called **exemplars**.

Exemplars

We already know that, compared to metrics, traces provide significantly more detail about what a system is doing. Due to the typical need to sample traces, though, they are not ideal for providing a complete, overarching view of your system in the way that metrics do. However, you can integrate the two systems through the use of Prometheus exemplars.

> **Enabling exemplars**
>
> At the time of writing, exemplars were still considered experimental features in Prometheus and must be enabled via the `--enable-feature=exemplar-storage` flag.

Exemplars help to balance the high-level system overview provided by metrics with the "drill-down" capabilities of traces. They work by attaching the latest trace ID to a time series data point when Prometheus collects it. Then, when the data is visualized (either in Prometheus or Grafana), a link to that trace can be presented to the user to quickly jump to a trace at that point in time.

Since traces are focused on being able to drill down into specific requests, the ability to jump from a spike in a metric to a trace during that time can help to further increase the usefulness of traces.

Summary

We've come to the end of our time together. I hope that you've enjoyed our time together as much as I have and that you've learned as much reading this book as I did writing it. Even with all the love I have for Prometheus, it is not the end-all-be-all of observability. It is but a piece in a much larger picture.

What precisely that picture looks like will vary from company to company and over time. As time progresses, new telemetry signals may become popular and be considered "core" observability signals. However, I feel confident in saying that metrics, logs, and traces are all here to stay. If you can build out your observability platform to cover all three of them, you'll be in a better position to monitor and observe your systems than the majority of people in our industry.

So, go forth with your new-found knowledge and use Prometheus as the cornerstone of your observability platform. Build it, scale it, tweak it, and tune it. Just know that there is always more waiting for you beyond Prometheus, too.

Further reading

To learn more about the topics that were covered in this chapter, take a look at the following resources:

- *OpenTelemetry specification*: `https://opentelemetry.io/docs/specs/otel/`
- *OpenTelemetry line protocol specification*: `https://opentelemetry.io/docs/specs/otlp/`
- *OpenTelemetry semantic conventions*: `https://opentelemetry.io/docs/specs/semconv/`
- *Recommended books on OpenTelemetry*:
  - Boten, Alex. *Cloud-Native Observability With OpenTelemetry: Learn to Gain Visibility Into Systems by Combining Tracing, Metrics, and Logging With OpenTelemetry*. Packt Publishing, 2022.
  - Parker, Austin, and Ted Young. *Learning OpenTelemetry: Setting up and Operating a Modern Observability System*. O'Reilly Media, 2024.
  - Majors, Charity, et al. *Observability Engineering*. O'Reilly Media, 2022.

Index

‹packt›

www.packtpub.com

Subscribe to our online digital library for full access to over 7,000 books and videos, as well as industry leading tools to help you plan your personal development and advance your career. For more information, please visit our website.

Why subscribe?

- Spend less time learning and more time coding with practical eBooks and Videos from over 4,000 industry professionals

- Improve your learning with Skill Plans built especially for you

- Get a free eBook or video every month

- Fully searchable for easy access to vital information

- Copy and paste, print, and bookmark content

Did you know that Packt offers eBook versions of every book published, with PDF and ePub files available? You can upgrade to the eBook version at packtpub.com and as a print book customer, you are entitled to a discount on the eBook copy. Get in touch with us at customercare@packtpub.com for more details.

At www.packtpub.com, you can also read a collection of free technical articles, sign up for a range of free newsletters, and receive exclusive discounts and offers on Packt books and eBooks.

Other Books You May Enjoy

If you enjoyed this book, you may be interested in these other books by Packt:

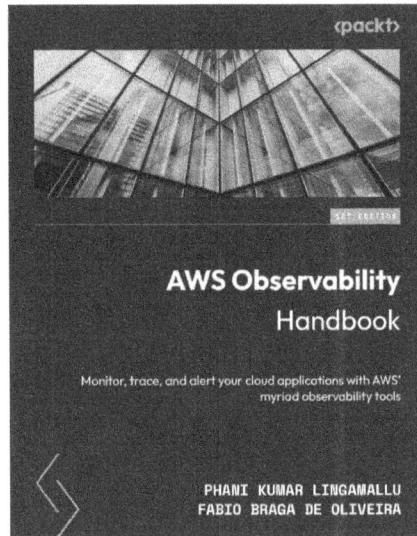

AWS Observability Handbook

Phani Kumar Lingamallu, Fabio Braga de Oliveira

ISBN: 978-1-80461-671-0

- Capture metrics from an EC2 instance and visualize them on a dashboard
- Conduct distributed tracing using AWS X-Ray
- Derive operational metrics and set up alerting using CloudWatch
- Achieve observability of containerized applications in ECS and EKS
- Explore the practical implementation of observability for AWS Lambda
- Observe your applications using Amazon managed Prometheus, Grafana, and OpenSearch services
- Gain insights into operational data using ML services on AWS
- Understand the role of observability in the cloud adoption framework

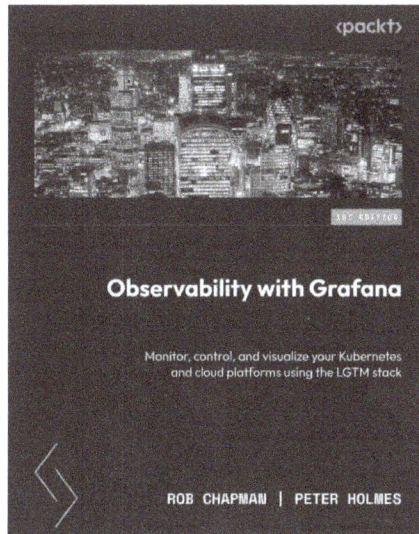

Observability with Grafana

Rob Chapman, Peter Holmes

ISBN: 978-1-80324-800-4

- Understand fundamentals of observability, logs, metrics, and distributed traces
- Find out how to instrument an application using Grafana and OpenTelemetry
- Collect data and monitor cloud, Linux, and Kubernetes platforms
- Build queries and visualizations using LogQL, PromQL, and TraceQL
- Manage incidents and alerts using AI-powered incident management
- Deploy and monitor CI/CD pipelines to automatically validate the desired results
- Take control of observability costs with powerful in-built features
- Architect and manage an observability platform using Grafana

Packt is searching for authors like you

If you're interested in becoming an author for Packt, please visit `authors.packtpub.com` and apply today. We have worked with thousands of developers and tech professionals, just like you, to help them share their insight with the global tech community. You can make a general application, apply for a specific hot topic that we are recruiting an author for, or submit your own idea.

Share Your Thoughts

Now you've finished *Mastering Prometheus*, we'd love to hear your thoughts! Scan the QR code below to go straight to the Amazon review page for this book and share your feedback or leave a review on the site that you purchased it from.

`https://packt.link/r/1-805-12566-4`

Your review is important to us and the tech community and will help us make sure we're delivering excellent quality content.

Download a free PDF copy of this book

Thanks for purchasing this book!

Do you like to read on the go but are unable to carry your print books everywhere?

Is your eBook purchase not compatible with the device of your choice?

Don't worry, now with every Packt book you get a DRM-free PDF version of that book at no cost.

Read anywhere, any place, on any device. Search, copy, and paste code from your favorite technical books directly into your application.

The perks don't stop there, you can get exclusive access to discounts, newsletters, and great free content in your inbox daily

Follow these simple steps to get the benefits:

1. Scan the QR code or visit the link below

https://packt.link/free-ebook/978-1-80512-566-2

2. Submit your proof of purchase
3. That's it! We'll send your free PDF and other benefits to your email directly

www.ingramcontent.com/pod-product-compliance
Lightning Source LLC
Chambersburg PA
CBHW081053220326

41598CB00038B/7085